Lecture Notes in Mathematics

continuation on page 213

Lecture Notes in Mathematics

Edited by A. Dold and B. Eckmann

748

Combinatorial Mathematics VI

Proceedings of the Sixth Australian
Conference on Combinatorial Mathematics,
Armidale, Australia, August 1978

Edited by
A. F. Horadam and W. D. Wallis

Springer-Verlag
Berlin Heidelberg New York 1979

Editors

A. F. Horadam
Department of Mathematics
University of New England
Armidale, N. S. W. 2351
Australia

W. D. Wallis
Department of Mathematics
University of Newcastle
Newcastle, N. S. W. 2308
Australia

AMS Subject Classifications (1970): 05 A 10, 05 A 99, 05 B 05, 05 B 15, 05 B 20, 05 B 35, 05 B 99, 05 C 05, 05 C 35, 05 C 99, 94 B 25

ISBN 978-3-540-09555-2 Springer-Verlag Berlin Heidelberg New York

2141/3140-543210

PREFACE

The sixth Australian conference on combinatorial mathematics was held at the University of New England, Armidale, in August 1978. The organising committee consisted of A.F. Horadam, B. Sims and W.D. Wallis. The names of the participants are listed overleaf.

We were fortunate enough to hear invited addresses by six distinguished combinatorists:

George Andrews (Pennsylvania State University);
Roger B. Eggleton (University of Newcastle);
Derek Holton (University of Melbourne);
Sheila Oates Macdonald (University of Queensland);
Don Row (University of Tasmania);
Ralph Stanton (University of Manitoba).

This volume contains texts of three of the invited addresses and most of the contributed talks.

The orginising committee wishes to thank all those people who helped with the running of the conference, in particular we are grateful to those who chaired the sessions, to those who refereed the contributions to this volume, and to the Vice-Chancellor and the Department of Mathematics of the University of New England.

A.F. Horadam
B. Sims
W.D. Wallis

TABLE OF CONTENTS

INVITED ADDRESSES

CONTRIBUTED PAPERS

SIXTH AUSTRALIAN CONFERENCE ON COMBINATORIAL MATHEMATICS

LIST OF PARTICIPANTS

G.E. ANDREWS   Department of Mathematics, Pennsylvania State University, University Park, Pa. 16802, U.S.A.

D. BILLINGTON   Department of Mathematics, University of Melbourne, Parkville, Victoria 3052

LYNDA BLATCH   c/- Department of Mathematics, University of New England, Armidale, N.S.W.   2351.

E.W. BOWEN   Department of Mathematics, University of New England, Armidale, N.S.W.   2351.

A. BRACE   School of Information Sciences, Canberra College of Advanced Education, P.O. Box 1, Belconnen, A.C.T. 2616.

D.R. BREACH   Department of Mathematics, University of Canterbury Christchurch, New Zealand.

J. CAMERON   Basser Department of Computer Science, University of Sydney, Sydney N.S.W.   2006.

G.J. CLARK   Department of Mathematics, High School, Narrabri, N.S.W.   2390.

P. EADES   Department of Computer Science, University of Queensland, St. Lucia, Queensland   4067.

R.B. EGGLETON   Department of Mathematics, University of Newcastle, Newcastle, N.S.W.   2308.

I. FAROUQI   Department of Mathematics, University of New England, Armidale, N.S.W.   2351.

L.R. FOULDS   Department of Mathematics, Massey University, Palmerston North, New Zealand.

D. GLYNN   Department of Pure Mathematics, University of Adelaide, Adelaide, South Australia   5001.

J. HAMMER   Department of Pure Mathematics, University of Sydney, Sydney, N.S.W.   2006.

P. HANLON   c/- Department of Mathematics, California Institute of Technology, Pasadena, California, U.S.A.

D. HOLTON   Department of Mathematics, University of Melbourne, Parkville, Victoria 3052

A.F. HORADAM   Department of Mathematics, University of New England, Armidale, N.S.W. 2351.

RUTH HUBBARD   Department of Mathematics, Queensland Institute of Technology, P.O. Box 246, North Quay, Queensland 4000.

N.A. JOHNSTON   c/- Department of Mathematics, University of New England, Armidale, N.S.W. 2351.

A.W. KINNERSLY   Department of Applied Mathematics, University of Sydney, Sydney, N.S.W. 2006.

C. KOHOUT   c/- Department of Mathematics and Computer Science, Royal Melbourne Institute of Technology, Melbourne, Victoria 3000.

R.P. LOH   Department of Applied Mathematics, University of Sydney, Sydney, N.S.W. 2006.

SHEILA OATES   Department of Mathematics, University of Queensland,
MACDONALD   St. Lucia, Queensland 4067.

K.L. McAVANEY   Division of Computing and Mathematics, Deakin University, P.O. Box 125, Belmont, Victoria 3216.

M. McKEE   c/- Mathematics Department, University of New England, Armidale, N.S.W. 2351.

ELIZABETH J.   Department of Mathematics, University of Queensland,
MORGAN   St. Lucia, Queensland 4067.

G.R. MORRIS   Department of Mathematics, University of New England, Armidale, N.S.W. 2351.

J.G. OXLEY   Department of Mathematics, Institute of Advanced Studies, Australian National University, P.O. Box 4, Canberra, A.C.T. 2600.

T. POLAK   c/- Department of Mathematics, University of New England, Armidale, N.S.W. 2351.

S. QUINN   Department of Mathematics, University of Newcastle, Newcastle, N.S.W. 2308.

A. RAHILLY   School of Applied Science, Gippsland Institute of Advanced Education, Churchill, Victoria 3842.

D.F. ROBINSON   Department of Mathematics, University of Canterbury, Christchurch, New Zealand.

R.W. ROBINSON   Department of Mathematics, University of Newcastle Newcastle, N.S.W. 2308.

C. RODGER   Department of Applied Mathematics, University of Sydney, Sydney, N.S.W. 2006.

D.G. ROGERS   c/- J.F.C. Kingman, Mathematical Institute, 24-29 St. Giles, Oxford OX1 3LB, England.

D.H. ROW — Department of Mathematics, University of Tasmania, G.P.O. Box 252C, Hobart, Tasmania 7001.

JENNIFER SEBERRY — Department of Applied Mathematics, University of Sydney, Sydney, N.S.W. 2006.

R.J. SIMPSON — Department of Mathematics, University of New England, Armidale, N.S.W. 2351.

B. SIMS — Department of Mathematics, University of New England, Armidale, N.S.W. 2351.

D.B. SKILLICORN — Department of Pure Mathematics, University of Sydney, Sydney, N.S.W. 2006.

KAYE STACEY — Department of Mathematics, Burwood State College, 221 Burwood Highway, Burwood, Victoria 3125.

R.G. STANTON — Department of Computer Science, University of Manitoba, Winnipeg, Manitoba, Canada.

ANNE PENFOLD STREET — Department of Mathematics, University of Queensland, St. Lucia, Queensland 4067.

DEBORAH J. STREET — 27 Exmouth Street, Toowong, Queensland 4066.

N.W. TAYLOR — Department of Mathematics, University of New England, Armidale, N.S.W. 2351.

W.D. WALLIS — Department of Mathematics, University of Newcastle, Newcastle, N.S.W. 2308.

J.E. WALTON — School of Cultural and Scientific Studies, Northern Rivers College of Advanced Education, P.O. Box 157, Lismore N.S.W. 2480.

B. WELLS — Department of Mathematics, Goulburn College of Advanced Education, Goulbourn, N.S.W. 2580.

C. WILKINS — Department of Mathematics, Sturt College of Advanced Education, Goulburn, N.S.W. 2580.

N.C. WORMALD — Department of Mathematics, University of Newcastle, Newcastle, N.S.W. 2308.

05C99

GRAPHIC SEQUENCES

R.B. EGGLETON AND D.A. HOLTON

The notion of degree sequence *(sometimes called* valence sequence*) of a graph is the basic theme in this discussion. Degree sequences are first considered in the context of simple graphs. Subsequently they are generalized to a wider setting which unifies various standard extensions of the definition of graph, and introduces further types of graph in a natural way.*

1. WHAT IS A DEGREE SEQUENCE?

The *degree* (or *valence*) of a vertex in a simple graph is the number of edges incident with that vertex. The *degree sequence* of a simple graph is the list of degrees of all the vertices, conventionally arranged monotonically and beginning with the maximum degree (Figure 1).

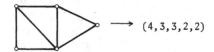

FIGURE 1 - A simple graph and its degree sequence

The passage from graph to degree sequence is straightforward, but problems are at once apparent if we try to pass in the reverse direction. For example, the sequence $(4,4,3,2,1)$ is not the degree sequence of any simple graph, whereas $(3,3,2,2,2)$ is the degree sequence of two nonisomorphic graphs (Figure 2).

FIGURE 2 - The two simple graphs with degree sequence $(3,3,2,2,2)$

2. WHEN IS A SEQUENCE GRAPHIC?

A finite sequence of non-negative integers is *graphic* if it is a permutation of the degree sequence of some graph. (Here we refer to simple graphs; the meaning of graphic will later be extended to other types of graph.)

PROBLEM 1. *Determine if a given finite sequence is graphic.*

FIRST ANSWER TO PROBLEM 1. (Havel [13]; Hakimi [10].) *Let* $\underline{d} = (d_0, d_1, d_2, \ldots)$

be a finite monotonic sequence of non-negative integers with maximum term d_0. *Then* $\underset{\sim}{d}$
is graphic precisely when $\underset{\sim}{d} - \underset{\sim}{e}$ *is graphic, the sequence* $\underset{\sim}{e}$ *being given by*

$$\underset{\sim}{e} = (d_0, \underbrace{1, \ldots, 1}_{d_0 \text{ terms}}) .$$

In this answer we adopt the convention that a linear combination of two sequences, such as $\underset{\sim}{d} - \underset{\sim}{e}$, is calculated termwise and if the sequences have different lengths, each unmatched term in the longer is treated as though it is matched with zero. The Havel-Hakimi answer to Problem 1 yields an algorithm for testing whether a given sequence is graphic. It reduces each graphic sequence to an all-zero sequence, and each non-graphic sequence to one with at least one negative term. Furthermore, if $\underset{\sim}{d}$ is graphic, the algorithm when applied to $\underset{\sim}{d}$ provides a means for constructing a *realization* of $\underset{\sim}{d}$, that is, a graph which has $\underset{\sim}{d}$ as its degree sequence. The construction uses the fact that if $\underset{\sim}{d}$ is graphic, it has a realization in which a vertex of degree d_0 is adjacent to d_0 other vertices of highest possible degree (Figure 3).

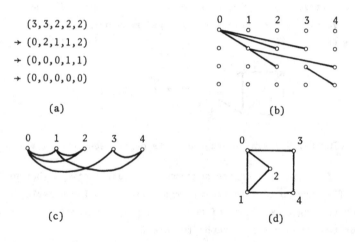

$$(3,3,2,2,2)$$
$$\rightarrow (0,2,1,1,2)$$
$$\rightarrow (0,0,0,1,1)$$
$$\rightarrow (0,0,0,0,0)$$

(a)

(b)

(c)

(d)

FIGURE 3 - (a) Algorithm showing that $(3,3,2,2,2)$ is graphic. (b) Corresponding realization, with separate copies of vertices for each step. (c) After identification of copies of vertices. (d) Rectilinear drawing of the realization.

To present another answer to Problem 1, some special notation is convenient. If $\underset{\sim}{d} = (d_0, d_1, d_2, \ldots)$ is a given sequence, its *kth initial segment* is the sequence $\alpha(\underset{\sim}{d}, k) = (d_0, d_1, \ldots, d_{k-1})$, its *kth final segment* is the sequence $\beta(\underset{\sim}{d}, k) = (\underbrace{0, \ldots, 0}_{k \text{ terms}}, d_k, d_{k+1}, \ldots)$, and its *kth level* is the sequence $\underset{\sim}{d}|k = \underset{\sim}{e} = (e_0, e_1, e_2, \ldots)$, where $e_i = \min \{d_i, k\}$. The *sum* of all terms in the sequence $\underset{\sim}{d}$ is denoted by $\Sigma \underset{\sim}{d}$.

SECOND ANSWER TO PROBLEM 1. (Erdös and Gallai [7]). *Let* $\underset{\sim}{d} = (d_0, d_1, d_2, \ldots)$
be a finite monotonic sequence of non-negative integers with maximum term d_0. *Then*
$\underset{\sim}{d}$ *is graphic just if*

(1) $\Sigma \underset{\sim}{d} \equiv 0 \pmod 2$, *and*

(2) $\Sigma\alpha(\underset{\sim}{d},k) \leqslant k(k-1) + \Sigma\beta(\underset{\sim}{d}|k,k)$, *for* $k = 1,2,3,\ldots$

The sufficiency of these conditions is not at all obvious, though their necessity is easily seen. Indeed, condition (1) corresponds to the fact that each edge of a simple graph is incident with exactly two vertices, while condition (2) corresponds to the fact that for each positive k, the induced subgraph on the k vertices i $(0 \leqslant i < k)$ of highest degree can contain at most $\binom{k}{2}$ edges, and the number of edges between this sub-graph and the vertex j (for $j \geqslant k$) is at most min $\{d_j,k\}$.

The Erdős-Gallai answer to Problem 1 can be formulated in a slightly different way using an idea of Fulkerson [8]. Given a sequence $\underset{\sim}{d} = (d_0,d_1,d_2,\ldots)$ of n non-negative integers, with $d_0 < n$, the *modified dual sequence* $\underset{\sim}{\hat{d}} = (\hat{d}_0,\hat{d}_1,\hat{d}_2,\ldots)$, also comprising n non-negative integers, is obtained from $\underset{\sim}{d}$ by a modified Ferrers diagram, as follows. In an $n \times n$ grid, crosses are entered in the principal diagonal sites, and d_i dots are entered into the available sites in row i $(0 \leqslant i < n)$, from left to right. Then $\hat{d}_i$ is the number of dots in column i $(0 \leqslant i < n)$. Clearly, the modified dual of $\underset{\sim}{\hat{d}}$ is $\underset{\sim}{d}$ itself (Figure 4). Note that if $d_0 \geqslant n$, then $\underset{\sim}{d}$ can still be treated if we first append $d_0 - n + 1$ zero terms.

	5	4	2	3	2	0	$- \underset{\sim}{\hat{d}}$
4	×	•	•	•	•		
4	•	×	•	•	•		
3	•	•	×	•			
2	•	•		×			
2	•	•			×		
1	•					×	

$\underset{\sim}{d} -$

FIGURE 4 - Modified dual of the sequence (4,4,3,2,2,1) is (5,4,2,3,2,0)

Using the modified dual sequence, the Erdős-Gallai answer to Problem 1 can be stated as follows:

Let $\underset{\sim}{d} = (d_0,d_1,d_2,\ldots)$ *be a finite monotonic sequence of non-negative integers with maximum term* d_0. *Then* $\underset{\sim}{d}$ *is graphic just if*

(1) $\Sigma \underset{\sim}{d} \equiv 0 \pmod 2$, *and*

(2) $\Sigma\alpha(\underset{\sim}{d},k) \leqslant \Sigma\alpha(\underset{\sim}{\hat{d}},k)$ *for* $k = 1,2,3,\ldots$

Let $\underset{\sim}{d} = (d_0,d_1,d_2,\ldots)$ be a finite monotonic sequence of non-negative integers with maximum term d_0. Then k is a *step* for $\underset{\sim}{d}$ if $d_k > d_{k+1}$. To test whether $\underset{\sim}{d}$ is graphic, it suffices to apply the Erdős-Gallai conditions at the steps (Eggleton [5]). Specifically, $\underset{\sim}{d}$ *is graphic just if*

4

(1) $\Sigma d \equiv 0 \pmod 2$,   *and*

(2) $\Sigma\alpha(\underline{d},k) \leqslant \Sigma\alpha(\hat{\underline{d}},k)$ *for each step* k *of* $\underline{d}$, *omitting the last if there are more than one.*

This result has particularly nice specializations to the cases where $\underline{d}$ has only one or two steps.

3.   HOW GRAPHIC IS A GRAPHIC SEQUENCE?

PROBLEM 2.   *Determine all distinct realizations of a given graphic sequence.*

Given any one realization of a graphic sequence, we can determine others by iterating a *switching* operation: if a,b,c,d are four vertices with adjacencies and non-adjacencies a~b, c~d, a$\not\sim$c, b$\not\sim$d, then switching replaces this configuration by a $\not\sim$ b, c $\not\sim$ d, a ~ c, b ~ d (Figure 5).

FIGURE 5 - The switching operation

Switching does not alter the degrees of a,b,c,d so the resultant graph is also a realization of the graphic sequence in question.   In fact, it solves the problem.

ANSWER TO PROBLEM 2.   (Eggleton [5]).   *Any realization of a graphic sequence can be obtained from any other by a sequence of switching operations.*

This result is closely related to work of Hakimi [10] on multigraphs.

To better understand how the realizations of a graphic sequence $\underline{d}$ are related, we introduce a graph R($\underline{d}$), called the *graph of realizations* of $\underline{d}$.   The vertices of R($\underline{d}$) are the distinct (that is, nonisomorphic) realizations of $\underline{d}$, and any two vertices G,H are adjacent if one can be obtained from the other by a single switching.   The answer to Problem 2 corresponds to the fact that R($\underline{d}$) is a connected graph.   The structure of R($\underline{d}$) is discussed in an accompanying paper [6].

4.   OTHER TYPES OF "GRAPH"

In addition to simple graphs, there are numerous variations to the notion of graph in the literature.   These include *multigraphs* (admitting multiple edges), *pseudographs* (admitting loops and multiple edges), *hypergraphs* (admitting edges incident with more than two vertices), *directed graphs* (with directions assigned to edges) and *directed pseudographs* (admitting loops and multiple directed edges).   Examples are given in Figure 6.   For each, the analogues of Problems 1 and 2 can be raised.

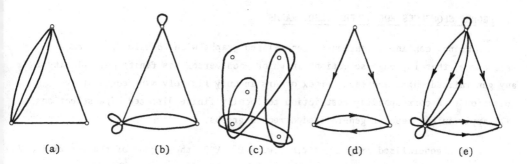

FIGURE 6 - Examples of (a) multigraph, (b) pseudograph, (c) hypergraph, (d) directed graph, (e) directed pseudograph

We shall now proceed to a generalization of degree sequences and an associated generalization of the notion of graph which includes the types just listed as particular cases.

5. GENERALIZED GRAPHS

Given any set S, let S* denote the set of all *sequences* (both finite and infinite) of elements from S. Two sequences $\underline{a}, \underline{b} \in$ S* are *equivalent* if there is a permutation of the index set of $\underline{a}$ which transforms it into $\underline{b}$. An equivalence class of sequences from S* is a *selection*. We let S° denote the set of all selections derived from S. In explicitly specifying a sequence we will use parentheses, while square brackets will be used for selections. Moreover, it is convenient to use an index notation for multiplicity of terms in a selection. To illustrate, $(2,3,2,0)$ and $(3,2,2,0)$ are equivalent sequences from the set of natural numbers $\mathbf{N} = \{0,1,2,...\}$. Both belong to the selection $[3,2,2,0]$, which can be written more concisely as $[3,2^2,0]$.

In what follows, we adopt the convention that V always denotes an initial segment, or all, of $\mathbf{N}$.

A *generalized edge* on V is any nonempty selection $e \in V°$. A *generalized graph* on V is an ordered pair G = (V,E), in which E is a selection of generalized edges on V, that is, $E \in V°°$. The elements of V are the *vertices* of G (Figure 7a).

A *generalized directed edge* on V is any nonempty sequence $\underline{e} \in$ V*. A *generalized directed graph* on V is an ordered pair G = (V,E), in which E is a selection of generalized directed edges on V, that is, $E \in V*°$. Again, the elements of V are the *vertices* of G (Figure 7b).

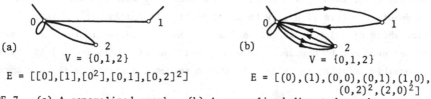

(a)
V = {0,1,2}
E = [[0],[1],[0^2],[0,1],[0,2]^2]

(b)
V = {0,1,2}
E = [(0),(1),(0,0),(0,1),(1,0), (0,2)^2,(2,0)^2]

FIGURE 7 - (a) A generalized graph. (b) A generalized directed graph

6. DEGREE SEQUENCES FOR GENERALIZED GRAPHS

We now confine attention to generalized graphs with *locally finite edge selections*, that is, edge selections in which each vertex has finite multiplicity in any generalized edge, and each vertex occurs in only finitely many edges of the selection. A corresponding restriction to locally finite directed edge selections will apply in the case of generalized directed graphs.

In a generalized graph $G = (V,E)$, with $E \in V^{\circ\circ}$, the *degree* of the vertex $i \in V$ is the sum of the multiplicities of i, taken over all edges in E. The *degree sequence* of G is the sequence $\underset{\sim}{d} \in \mathbb{N}^*$, with index set V, such that $\underset{\sim}{d}(i)$, the term with index i, is equal to the degree of i in G. For example, the generalized graph in Figure 7a has degree sequence $(6,2,2)$.

Similarly, for a generalized directed graph $G = (V,E)$, with $E \in V^{*\circ}$, the *degree* of the vertex $i \in V$ is a sequence $\underset{\sim}{d}(i) \in \mathbb{N}^*$, of length equal to the supremum of lengths of directed edges in E, and such that $\underset{\sim}{d}(i,j)$, the term in $\underset{\sim}{d}(i)$ with index j, is equal to the sum of multiplicities of occurrence of i as the term with index j in directed edges in E. The *degree sequence* of G is $\underset{\sim}{d} \in \mathbb{N}^{**}$, a sequence of sequences, with index set V, and such that $\underset{\sim}{d}(i)$, the term with index i, is the degree of i in G. For example, the generalized directed graph in Figure 7b has degree sequence $((5,4),(2,1),(2,2))$.

A sequence $\underset{\sim}{d} \in \mathbb{N}^*$ is *graphic* if it is the degree sequence of some generalized graph.

PROBLEM 3.   *Which sequences $\underset{\sim}{d} \in \mathbb{N}^*$ are graphic?*

ANSWER TO PROBLEM 3.   *Every $\underset{\sim}{d} \in \mathbb{N}^*$ is graphic.*

Given $\underset{\sim}{d} = (d_0,d_1,d_2,\ldots) \in \mathbb{N}^*$, a trivial realization of $\underset{\sim}{d}$ is provided by $G = (V,E)$, where V is the index set of $\underset{\sim}{d}$ and $E = [[0]^{d_0},[1]^{d_1},[2]^{d_2},\ldots]$. Another obvious realization $G' = (V,E')$ involves just one generalized edge: $E' = [[0^{d_0},1^{d_1},2^{d_2},\ldots]]$.

A sequence $\underset{\sim}{d} \in \mathbb{N}^{**}$ is *graphic* if it is the degree sequence of some generalized directed graph.

PROBLEM 4.   *Which sequences $\underset{\sim}{d} \in \mathbb{N}^{**}$ are graphic?*

To answer Problem 4 we need several definitions. A sequence $\underset{\sim}{d} \in \mathbb{N}^{**}$ is *uniform* if all terms of $\underset{\sim}{d}$, themselves sequences in $\mathbb{N}^*$, have the same index set I. If V is the index set of $\underset{\sim}{d}$, then $\underset{\sim}{d}(i)$ will denote the term of $\underset{\sim}{d}$ with index $i \in V$, and $\underset{\sim}{d}(i,j)$ will denote the term of $\underset{\sim}{d}(i)$ with index $j \in I$. If for each $i \in V$ there is a largest j for which $\underset{\sim}{d}(i,j) > 0$, we call $\underset{\sim}{d}$ *locally finite*. (However, the set of such largest indices need not be bounded.) The *component sum* of $\underset{\sim}{d}$ with *index* j is the sum $\sum_{i \in V} \underset{\sim}{d}(i,j)$, possibly infinite in value. A uniform sequence d satisfies the *monotone* *component sum condition* if the component sums satisfy $\sum_{i \in V} d(i,j) \geq \sum_{i \in V} d(i,k)$ for every j,k $\in$ I with j < k. It is rather straightforward to prove the following result.

ANSWER TO PROBLEM 4. *Generalized directed graphic sequences corresponding to locally finite generalized directed graphs are just those sequences* $\underset{\sim}{d} \in \mathbb{N}^{**}$ *which are uniform and locally finite, and satisfy the monotone component sum condition.*

7. TYPES OF GENERALIZED GRAPH

To discuss the various types of generalized graph, we first classify edges. Two generalized edges on a vertex set V are *similar* if there is a permutation of V which transforms one into the other. From any similarity class of edges on V, we select as canonical representative the first member of the class under the lexiographic ordering inherited from $\mathbb{N}$. Thus, the canonical representatives of similarity classes of generalized edges with at most three vertices are:

$$[0],[0^2],[0,1],[0^3],[0^2,1],[0,1,2].$$

Our terminology for a member of one of these classes is *spine, loop, edge, sphere, cone* and *triangle*, respectively. It is clear how this kind of classification can be adapted to generalized directed edges, so we omit the details.

We now confine attention to locally finite generalized graphs in which all edges are finite (though the number of edges may be infinite). A subset of these generalized graphs is a *type* if its members are all those in which the multiplicity of each generalized edge satisfies a specific upper bound (possibly infinite) which depends only on the similarity class of the edge.

A type can be specified by listing the canonical representatives of the similarity classes of edges, each with an exponent corresponding to the bound on multiplicity for edges of that class. (Usually it is convenient to omit representatives of classes with 0 as upper bound, and to omit the index 1 if it is the upper bound.) To illustrate, *multigraphs* have type $[0,1]^\infty$, *simple graphs* have type $[0,1]$, and *pseudographs* have type $[0^2]^\infty,[0,1]^\infty$.

Alternatively, a type can be specified by its *type sequence* τ, defined as follows. Let $\underset{\sim}{c}$ denote the sequence of canonical representatives of similarity classes of generalized edges, in which edges with fewer vertices (counted according to multiplicity) precede those with more vertices, and edges with the same number of vertices occur in lexicographic order. (The first six terms of $\underset{\sim}{c}$ were listed in order when canonical representatives of similarity classes were discussed above.) The type sequence τ of a particular type has each term $\tau(i)$ equal to the bound on multiplicity of edges similar to $\underset{\sim}{c}(i)$. (Usually it is convenient to omit all terms of τ beyond the last nonzero term.) Thus, *multigraphs* have type sequence $(0,0,\infty)$, *simple graphs* have type sequence $(0,0,1)$, and *pseudographs* have type sequence $(0,\infty,\infty)$.

Similar definitions and notation apply in the case of generalized directed graphs. For example, *directed multigraphs* have (directed) type $(0,1)^\infty$ and *directed pseudographs* have (directed) type $(0,0)^\infty, (0,1)^\infty$. The sequence $\underline{c}'$ of canonical representatives of similarity classes of generalized directed edges begins

$$\underline{c}' = \big((0),(0,0),(0,1),(0,0,0),(0,0,1),(0,1,0),(0,1,1),(0,1,2)\ldots\big)$$

and directed type sequences are defined as before, but with the convention that when listed explicitly they are enclosed in angle brackets rather than parentheses. Thus, *directed multigraphs* have (directed) type sequence $<0,0,\infty>$.

A whole range of problems can now be posed, for each possible type sequence τ.

PROBLEM 5. *Which sequences $\underline{d} \in \mathbb{N}^*$ are graphic sequences of type τ?*

PROBLEM 6. *Which sequences $\underline{d} \in \mathbb{N}^{**}$ are graphic sequences of directed type τ?*

PROBLEM 7. *Find a good algorithm for producing a type τ realization of $\underline{d}$, if such a realization exists.*

PROBLEM 8. *What operation on type τ realizations of $\underline{d}$ can be used to define adjacency of realizations so that the connected components of $R(\underline{d};\tau)$ are as large as possible?*

PROBLEM 9. *What is the structure of $R(\underline{d};\tau)$?*

Some of the answers which have been obtained will now be mentioned briefly.

Solutions to Problem 5 have been obtained for a number of types. When $\tau = (0,0,1)$, they are precisely the answers to Problem 1; when $\tau = (0,0,\infty)$, the solution was apparently first obtained by Senior [16], though a number of independent rediscoveries have been made; when $\tau = (0,m,n)$, and in particular when $\tau = (0,0,n)$, the solution was given by Chen [3]. Also, the solution when $\tau = (0,\infty,\infty)$ was given by Senior [16]. The trivial problems when $\tau = (\infty)$ and $\tau = \underline{1} = (1,1,1,\ldots)$ are solved by our constructions to answer Problem 3.

Solutions to Problem 6, the directed analogue of Problem 5, have also been obtained for a number of types. When $\tau = <0,1,1>$, the answer is due to Gale [9] and Ryser [15]. The solution when $\tau = <0,n,n>$ was obtained by Berge [2]. The solution when $\tau = <0,0,1>$ was given by Fulkerson [8], and simplified by Chen [3], who also gave solutions when $\tau = <0,\infty,\infty>$, $\tau = <0,m,n>$ and in particular $\tau = <0,0,n>$.

Problem 7 when $\tau = (0,0,1)$ is solved by the Havel-Hakimi answer to Problem 1. When $\tau = <0,m,n>$, the solution was given by Chen [3], and was also adapted to deal with $\tau = (0,0,n)$.

Problem 8 when $\tau = (0,0,\infty)$ was solved by Hakimi [10]. When $\tau = (0,0,1)$, the solution is given by the answer to Problem 2. When $\tau = <0,m,n>$, the solution was obtained by Hakimi [12] and Chen [3].

A number of results can be regarded as relevant to Problem 9. For example, Hakimi [10] characterized all cases in which some vertex of $R(\underline{d};(0,0,\infty))$ is (i) connected, or (ii) nonseparable, or (iii) connected and separable. Menon [14] determined when some vertex of $R(\underline{d};<0,m,n>)$ is a tree, and Hakimi [12] and Chen [3] determined when some vertex is connected. Senior [16] characterized all cases in which $R(\underline{d};(0,0,\infty))$ has only one vertex which is connected, and Bäbler [1] and Hakimi [11] showed when there is only one vertex altogether. In the accompanying paper [6] we examine the structure of $R(\underline{d};(0,\infty,\infty))$.

A detailed discussion of many of these results is given in Chen [4].

REFERENCES

[1]   F. Bäbler, Über eine spezielle Klasse Euler'sche Graphen, *Comment. Math. Helv.*, 27 (1953), 81-100.

[2]   C. Berge, *Theory of Graphs and Its Applications* (Methuen, London, 1962).

[3]   W.K. Chen, On the realization of a (p,s)-digraph with prescribed degrees, *J. Franklin Inst.*, 281 (1966), 406-422.

[4]   W.K. Chen, *Applied Graph Theory* (North Holland, Amsterdam, 2nd revised edn., 1976).

[5]   R.B. Eggleton, Graphic sequences and graphic polynomials: a report, in *Infinite and Finite Sets*, Vol. 1, ed. A. Hajnal *et al*, Colloq. Math. Soc. J. Bolyai 10, (North Holland, Amsterdam, 1975), 385-392.

[6]   R.B. Eggleton and D.A. Holton, The graph of type $(0,\infty,\infty)$ realizations of a graphic sequence, *this volume*.

[7]   P. Erdös and T. Gallai, Graphs with prescribed degrees of vertices (in Hungarian), *Mat. Lapok*, 11 (1960), 264-274.

[8]   D.R. Fulkerson, Zero-one matrices with zero trace, *Pacific J. Math.*, 10 (1960), 831-836.

[9]   D. Gale, A theorem on flows in networks, *Pacific J. Math.*, 7 (1957), 1073-1082.

[10]   S.L. Hakimi, On realizability of a set of integers as degrees of the vertices of a linear graph I, *J. Soc. Indust. Appl. Math.*, 10 (1962), 496-506.

[11]   S.L. Hakimi, On realizability of a set of integers as degrees of the vertices of a linear graph II: Uniqueness, *J. Soc. Indust. Appl. Math.*, 11 (1963), 135-147.

[12]   S.L. Hakimi, On the degrees of the vertices of a directed graph, *J. Franklin Inst.*, 279 (1965), 290-308.

[13]   V. Havel, A remark on the existence of finite graphs (in Czech), *Časopis Pěst. Mat.*, 80 (1955), 477-480.

[14]   V.V. Menon, On the existence of trees with given degrees, *Sankhyā*, Ser. A., 26 (1964), 63-68.

[15]   H.J. Ryser, Combinatorial properties of matrices of zeros and ones, *Can. J. Math.*, 9 (1957), 371-377.

[16]   J.K. Senior, Partitions and their representative graphs, *Amer. J. Math.*, 73 (1951), 663-689.

Department of Mathematics,
University of Newcastle,
New South Wales, 2308,
Australia.

Department of Mathematics,
University of Melbourne,
Parkville, Vic.   3052,
Australia.

05 - 02
94 A 10
05 B 99
03 B 45

COMBINATORICS - A BRANCH OF GROUP THEORY?

SHEILA OATES MACDONALD

After a brief survey of the interaction between combinatorics and group theory, three illustrations of the application of group theory to combinatorial problems are given.

1. INTRODUCTION

The relationship between combinatorics and group theory may be divided into three parts - combinatorics as a tool in group theory, group theory as a tool in combinatorics, and finally situations where the relationship between the two is such that it is impossible to determine which is the master and which the slave.

In the first category I would place the theory of graphs of groups, an exposition of which was given by Imrich [8], and also the construction of new simple groups related to automorphism groups of combinatorial objects, perhaps the most notable example of this is the automorphism group of the Leech lattice, which is based on the Golay code,Leech [9,10]. This group was shown by Conway [2] to have as sections three new simple groups as well as several of the sporadic simple groups previously discovered.

In the third category I would place stability theory of graphs, and Cayley graphs of groups. But it is from the second category that I have chosen my illustrations. Here there are many examples - the use of starters and adders in the construction of Room squares([20], part 2), difference sets in the construction of Hadamard matrices ([20], part 4), the use of extensions of automorphism groups of block designs to extend the designs, permutational decoding, sum free sets in groups, and colourings of tessellations. It is on the last three that I propose to elaborate.

2. PERMUTATIONAL DECODING

The basic connection between codes and group theory is, of course, obvious, since linear codes are vector subspaces and so, in particular, are abelian groups. However, here I shall consider the use of the automorphism group of a code in its decoding. (The ideas in this section come from McWilliams [15].)

We will consider an e-error-correcting (n,k)-code over $GF[2]$ where the first k

positions contain the information (and can take all possible values), the remaining n-k consist of parity checks uniquely determined by the entries in the first
k places, and any two code words are at least a distance of 2e+1 apart. (Here
distance is the Hamming distance - the number of places in which the code words
differ.)

Let β be a binary sequence of length k and let $m(\beta)$ denote the code word
determined by β. Let α be an arbitrary sequence of length n and $\bar{\alpha}$ the sequence of its
first k letters; then α is a code word if and only if $m(\bar{\alpha}) = \alpha$.

Suppose now that α is a received word. If there are no errors in the first k
places of α then $\alpha_0 = m(\bar{\alpha})$ is the corrected version of α, and must be of distance $\leq$ e
from α. On the other hand, if there are errors in the first k places, then α_0 is not
the corrected version of α and so its distance from α is greater than e. Let π be an
element of the automorphism group of the code which permutes the letters; then the
distance between two words α and β will be the same as the distance between $\alpha\pi$ and $\beta\pi$.
Hence if we can find π such that the distance between $\alpha\pi$ and $m(\overline{\alpha\pi})$ is less than e then
the correct version of α is $m(\overline{\alpha\pi})\pi^{-1}$. Clearly π must have the property that it moves
any error out of the first k places, so the problem is to find a minimal set of auto-
morphisms of the code which are guaranteed to move any error out of the first k places,
and many open questions remain in this area (see McWilliams and Sloan [16] Ch.16, §9).

I shall end this section with an illustration of the use of this technique to
decode the Hamming (7,4)-code (Note that this is not the most efficient decoding
method for this code.) The sixteen code words are:

$$
\begin{array}{ccccccc}
0 & 0 & 0 & 0 & 0 & 0 & 0 \\
0 & 1 & 1 & 0 & 1 & 0 & 0 \\
0 & 0 & 1 & 1 & 0 & 1 & 0 \\
0 & 0 & 0 & 1 & 1 & 0 & 1 \\
1 & 0 & 0 & 0 & 1 & 1 & 0 \\
0 & 1 & 0 & 0 & 0 & 1 & 1 \\
1 & 0 & 1 & 0 & 0 & 0 & 1 \\
1 & 1 & 0 & 1 & 0 & 0 & 0 \\
1 & 1 & 1 & 1 & 1 & 1 & 1 \\
1 & 0 & 0 & 1 & 0 & 1 & 1 \\
1 & 1 & 0 & 0 & 1 & 0 & 1 \\
1 & 1 & 1 & 0 & 0 & 1 & 0 \\
0 & 1 & 1 & 1 & 0 & 0 & 1 \\
1 & 0 & 1 & 1 & 1 & 0 & 0 \\
0 & 1 & 0 & 1 & 1 & 1 & 0 \\
0 & 0 & 1 & 0 & 1 & 1 & 1 \\
\end{array}
$$

This is a single-error-correcting code and is invariant under cyclic permutations of
the letters, which clearly will suffice to move any error out of the first four places,

(in fact, the set 1, π^3, π^6 will suffice, where π denotes the cyclic permutation moving each letter one place to the right.) Suppose the received word is α = 1001101, then $\alpha\pi^3$ = 1011001, $\alpha\pi^6$ = 0011011, m($\overline{\alpha}$) = 1001011, m($\overline{\alpha\pi^3}$) = 1011100, m($\overline{\alpha\pi^6}$) = 0011010, which are distances 2,2,1 from α, $\alpha\pi^3$, $\alpha\pi^6$ respectively. Hence 0011010 is the corrected version of $\alpha\pi^6$, and hence 0001101 is the corrected version of α.

3.   SUMFREE SETS AND RAMSEY NUMBERS

In this section I want to consider the application of sumfree sets to the determination of certain Ramsey numbers. The basic reference for most of the ideas here is part 3 of Wallis, Street and Wallis [20].

Ramsey's theorem on subsets of a finite set states: for any positive integers $m_1, \ldots, m_r$; t with $k_i \geq r$, there exists a positive integer R = R($k_1, \ldots, k_r$;t) such that for any s $\geq$ R and any partition of the set of t-subsets of an s-set S into disjoint classes $A_1, \ldots, A_r$ there exists an m_i-subset M_i of S all of whose t-subsets belong to A_i.

The case where all the m_i are equal to 3 and t is 2 has an interpretation as a question about edge colourings of a complete graph. Indeed, if

$$R = R(\underbrace{3, \ldots, 3}_{r};2) = R_r(3;2),$$

then K_R is the smallest complete graph such that any edge colouring using r colours forces the appearance of a monochromatic triangle. (A colouring with no monochromatic triangle will be called *proper*.) Some of the results about the sizes of the $R_r(3;2)$, in particular lower bounds, have been obtained using sumfree sets in groups.

Definition. A subset S of a group G (written additively) is called *sumfree* if (S+S) $\cap$ S = $\emptyset$. It is *symmetric* if S = -S.

Let G be a group of order n with its elements ordered arbitrarily. Suppose the non-zero elements of G can be partitioned into r sumfree sets; then a proper r-colouring of K_n is obtained by labelling the vertices of the graph with the elements of G and colouring the edge joining the vertices g and h with colour c_i if g > h and g-h $\in$ S_i. To see that this is a proper colouring, consider a triangle with vertices g, h and k, where g > h > k then, since g-k = g-h+h-k, if edges (g,h) and (h,k) have the same colour, edge (g,k) must have a different one. Note that if the sumfree sets can be chosen to be symmetric, then the ordering is unnecessary, since g-h and h-g will belong to the same sumfree set. A simple example of such a colouring is the 2-colouring of K_5 shown in Figure 1. Here the vertices are labelled with the elements of the cyclic group of order 5, and the requisite sumfree sets are {1,4} and {2,3} (these are, in fact, both symmetric, so no ordering is needed.)

A simple argument, based on the fact that in any 2-colouring of K_6 three of the edges at a vertex must receive the same colour, shows that K_6 has no proper 2-colouring, so that $R_2(3,2) = 6$.

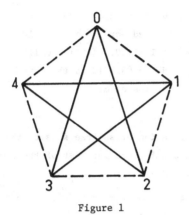

Figure 1

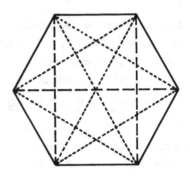

Figure 2

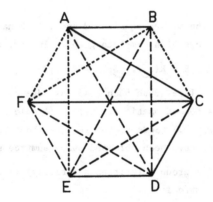

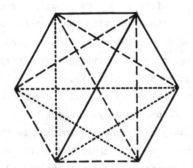
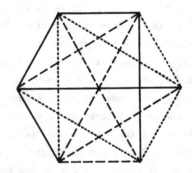

Figure 3

A generalisation of the argument used for K_6 shows that

$$R_{r+1}(3;2) \leq (r+1)(R_r(3;2)-1) + 2 ,$$

so $R_3(3;2) \leq 17$, and a proper 3-colouring of K_{16} exists (indeed, there are two non-isomorphic such colourings, arising respectively from sumfree partitions of $\mathbb{Z}_2 + \mathbb{Z}_2 + \mathbb{Z}_2 + \mathbb{Z}_2$ and $\mathbb{Z}_4 + \mathbb{Z}_4$) so that $R(3;2) = 17$.

The situation now deteriorates. The above formula gives $R_4(3,2) \leq 4.16+2 = 66$, and, in fact, Folkman [3] and Whitehead [21] have shown, by consideration of the incidence matrix, that $R_4(3,2) \leq 65$. A lower bound of 50 was obtained by Whitehead [21] by partitioning the non zero elements of $\mathbb{Z}_7 + \mathbb{Z}_7$ into four sumfree sets. This was improved by Chung [1] to 51, using incidence matrices. There remains the gap between 51 and 65 to be filled. There are many groups of orders in this range, particularly of order 64. Perhaps one worth mentioning is A_5 (of order 60). Street [18] has shown that there is a sumfree set of order 18 in A_5, but so far all efforts to partition the non-identity elements of A_5 into four sumfree sets have proved unavailing.

Although the only proper 2-colouring of K_5 and the two proper 3-colourings of K_{16} all arise from sumfree sets in groups, if we consider the other end of the scale, the 330 proper 3-colourings of K_6 (Heinrich [6]), we find plenty of colourings that do not arise from sumfree sets in groups of order 6. Indeed, she has shown, [7], that one of these (Figure 2) cannot even be obtained by embedding K_6 into any proper 3-colouring of the complete graphs of order up to 16 obtained from sumfree partitions of groups of appropriate orders. There are also several other colourings of K_6 for which no such embedding is known. However, Street [19] has shown that for sufficiently large m and s any proper r-colouring of K_n can be embedded in a proper s-colouring of K_m arising from a sumfree partition of the cyclic group of order m.

In [11] I endeavoured to complicate the issue by considering sumfree sets in loops. Of course, the lack of associativity causes additional problems, but I was able to obtain a loop of order six with sumfree sets which yielded the colouring of Figure 2, and also loops which worked for some of the colourings listed in [6] as "no imbedding known". However, it is very easy to prove that other such colourings, for instance the three shown in Figure 3, cannot arise from a partition of the non-ident-ity elements of a loop of order six into sum-free sets. The proofs are based on the observation that if a vertex is incident with three edges of the same colour, then the sumfree set corresponding to that colour must contain at least two elements (since otherwise we would have a-b = a-c, or b-a = c-a, which is impossible in a loop, which-ever definition of "-" we choose to use). This immediately rules out the third colour-ing since to each colour there is a vertex incident with three edges of that colour, and it is, of course, impossible to partition a set of size five into three sets of size at least two. For the other two colourings it can be shown that there is no ordering of the elements of the loop that will work. For instance, consider the first one. Both the solid edges and the dashed edges must correspond to sumfree sets of size at least two (and so of size precisely two), so the dotted edges correspond to a sumfree set of size one, hence the vertices along the dotted path must be ordered either

$$C > B > F > A > E \quad \text{or} \quad E > A > F > B > C .$$

Also C is linked to vertices A, D, and F by solid edges so cannot be greater than all three, or less than all three (since otherwise we would have two of C-A, C-D, C-F, or two of A-C, D-C, F-C equal). Similarly D cannot be greater than all three of A, B, F, nor less than all three. But now there is no way of fitting D into either of the above orderings which satisfies all these conditions. The argument for the second colouring is similar.

Thus loops are certainly not the answer to all our problems, and I would hesitate to suggest that anyone turn to loops to resolve the $R_4(3;2)$ question - at least not until the available groups have been exhausted.

4.   COLOURINGS OF TESSELLATIONS

In several papers recently, (see [5] for a full list) Grünbaum and Shephard have classified various tessellations and, once one has a tessellation, it is natural to consider colourings. As with the colourings of the fundamental regions of the crystallographic groups which Anne Street and I have considered in [12], [13] and [14], (see also Roth [17]). I shall consider only colourings such that the symmetries of the pattern permute the colours. Thus we are seeking certain permutation representations of the automorphism group of the tessellation.

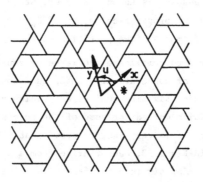

Figure 4

In the case of the crystallographic groups the group was transitive on the fundamental regions, and so the permutation representation, being transitive, arises from a representation on the cosets of a subgroup, and, since only the identity element fixes a fundamental region, there is no restriction on the choice of subgroup. Neither of these properties need hold for a general tiling, as can be seen from the tessellat-

$H_t = H_h = G$

$H_t = \langle u^2, x, y \rangle$
$H_h = G$

$H_t = G$
$H_h = \langle u, x^3, xy \rangle$

$H_t = \langle u^2, x, y \rangle$
$H_h = \langle u, x^3, xy \rangle$

$H_t = G$
$H_h = \langle u, x^2, y^2 \rangle$

$H_t = \langle u^2, x, y \rangle$
$H_h = \langle u, x^2, y^2 \rangle$

Figure 5(i)

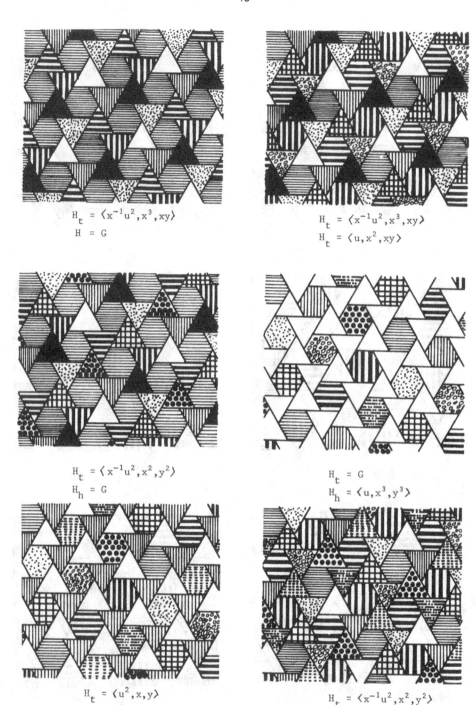

$$H_t = \langle x^{-1}u^2, x^3, xy \rangle$$
$$H = G$$

$$H_t = \langle x^{-1}u^2, x^3, xy \rangle$$
$$H_t = \langle u, x^2, xy \rangle$$

$$H_t = \langle x^{-1}u^2, x^2, y^2 \rangle$$
$$H_h = G$$

$$H_t = G$$
$$H_h = \langle u, x^3, y^3 \rangle$$

$$H_t = \langle u^2, x, y \rangle$$
$$H_h = \langle u, x^3, y^3 \rangle$$

$$H_t = \langle x^{-1}u^2, x^2, y^2 \rangle$$
$$H_h = \langle u, x^3, xy \rangle$$

Figure 5(ii)

ion illustrated in Figure 4 (see [4] for others of this type). Here, since the tiles are not congruent, the symmetry group G certainly cannot be transitive on the tiles, though it is transitive on the hexagons and on the triangles. Also there are non-trivial symmetries fixing tiles. Indeed, if u is the rotation through $\pi/3$ about the centre of a hexagon, and x and y are translations, as shown on Figure 4 (so that $G = gp\{u,x,y \mid u^6 = [x,y] = 1, x^u = y, y^u = x^{-1}y\}$), then the corresponding hexagon is fixed by $gp\{u\}$ and the marked triangle by $gp\{x^{-1}u^2\}$. If the permutation representation on the triangles corresponds to the subgroup H_t and that on the hexagons to H_h, then $x u^2 \in H_t$ and $u \in H_h$. Since the only subgroup containing both these elements is G itself, the representations on the triangles and hexagons are distinct (except for the identity representation) so different colours must be used. Colourings in two to eleven colours and the corresponding H_t and H_u are shown in Figure 5.

[It may be of interest to note that in general, if $T = gp\{x,y\}$ then $T/H_t \cap T$ and $T/H_h \cap T$ are of the form $\mathbb{Z}_r + \mathbb{Z}_r$, with the exception of the three colourings for the hexagons and the six colourings for the triangles, where these factor groups are $\mathbb{Z}_3$.]

REFERENCES

[1]   Fan Rong K. Chung, On the Ramsey Numbers N(3,...,3;2), *Discrete Math.* 5 (1973), 317-321.

[2]   J.H. Conway, A group of order 8, 315, 553, 613, 086, 720, 000. *Bull. London Math. Soc.* 1 (1969), 79-86.

[3]   Jon Folkman, Notes on the Ramsey number N(3,3,3,3), Manuscript, Rand Corporation, Santa Monica, California, 1967.

[4]   Branko Grünbaum and G.C. Shephard, Tilings by regular polygons, *Math. Mag.* 50 (1977), 227-247.

[5]   Branko Grünbaum and G.C. Shephard, Incidence symbols and their applications, *Proc. Symposium on relations between Combinatorics and other parts of Mathematics,* Columbus, Ohio, 1978 (to appear).

[6]   K. Heinrich, Proper 3-colourings of K_6, *Ars Combinatoria* 1 (1976), 191-213.

[7]   K. Heinrich, A non-imbeddable proper colouring, *Combinatorial Mathematics IV, Proc. Fourth Australian Conf. Lecture Notes in Math. 560,* 93-115, (Springer-Verlag, Berlin, Heidelberg, New York, 1976).

[8]   Wilfred Imrich, Subgroup theorems and graphs, *Combinatorial Mathematics V, Proc. Fifth Australian Conf. Lecture Notes in Math. 622,* 1-27, (Springer-Verlag, Berlin, Heidelberg, New York, 1977).

[9]   John Leech, Some sphere packings in higher space, *Canad. J. Math.* 16 (1964), 657-682.

[10] John Leech, Notes on sphere packings, *Canad. J. Math.* 19 (1977), 251-267.

[11]   Sheila Oates Macdonald, Sum-free sets in loops, *Combinatorial Mathematics V,*
       Proc. Fifth Australian Conf. Lecture Notes in Math. 622, 141-147, (Springer-
       Verlag, Berlin, Heidelberg, New York, 1977).

[12]   Sheila Oates Macdonald and Anne Penfold Street, On crystallographic colour
       groups, *Combinatorial Mathematics IV, Proc. Fourth Australian Conf. Lecture*
       Notes in Math. 560, 149-157, (Springer-Verlag, Berlin, Heidelberg, New York,
       1976).

[13]   Sheila Oates Macdonald and Anne Penfold Street, The seven friezes and how to
       colour them, *Utilitas Math.* 13 (1978), 271-292.

[14]   Sheila Oates Macdonald and Anne Penfold Street, The analysis of colour symmetry,
       Combinatorial Mathematics VI, Proc. International Conf. (Canberra 1977),
       Lecture Notes in Math. (Springer-Verlag, Berlin, Heidelberg, New York,
       to appear).

[15]   F.J. McWilliams, Permutation decoding of systematic codes, *Bell System Tech. J.*
       43 (1964), 485-505.

[16]   F.J. McWilliams and N.J.A. Sloane, *The theory of error-correcting codes,*
       (North-Holland, Amsterdam, New York, Oxford, 1977).

[17]   Richard L. Roth, Colour symmetry and group theory (preprint, 1978).

[18]   Anne Penfold Street, A maximal sum-free set in A_5, *Utilitas Math.* 5 (1974),
       85-91.

[19]   Anne Penfold Street, Embedding proper colourings, *Combinatorial Mathematics IV,*
       Proc. Fourth Australian Conf. Lecture Notes in Math. 560, 240-245, (Springer-
       Verlag, Berlin, Heidelberg, New York, 1976).

[20]   W.D. Wallis, Anne Penfold Street and Jennifer Seberry Wallis, *Combinatorics :*
       Room squares, sum-free sets and Hadamard matrices. Lecture Notes in Math. 292,
       (Springer-Verlag, Berlin, Heidelberg, New York, 1972).

[21]   Earl Glen Whitehead, Jr. The Ramsey number N(3,...,3;2), *Discrete Math.* 4
       (1973), 389-396.

Department of Mathematics
University of Queensland
St. Lucia, 4067
Queensland
Australia

THE CURRENT STATUS OF THE GENERALISED MOORE GRAPH PROBLEM

B.D. McKAY AND R.G. STANTON

1. INTRODUCTION AND BACKGROUND

The concept of a generalised Moore graph has been developed by Cerf, Cowan, Mullin, Stanton, in a series of paper [2]-[7]. Basically, the problem arose in connection with the topological design of computer communication networks; in Figure 1, we reproduce from [3] the graph of $P(N,V)$ against N, where N is the number of nodes in a regular graph of valence V, $P(N,V)$ is the average path length in the graph, and $P(N,V)$ is minimal. It is clear from the figure that $P(N,V)$ grows logarithmically for $V \geqslant 3$, and we have restricted our published discussions to the case $V = 3$, which is typical, economical and convenient.

2. REPHRASING THE PROBLEM

To rephrase the problem in purely graph-theoretic terms, we look at one particular node, which we call the root node R. Recall that we are dealing with valence 3; then there are to be 3 nodes at unit distance from R, 6 at distance 2 from R, 12 at distance 3 from R, etc. We determine k such that

$$1 + 3 + 6 + 12 + \ldots + 3.2^{k-2} < N,$$
$$1 + 3 + 6 + 12 + \ldots + 3.2^{k-1} \geqslant N.$$

This ensures that all nodes are packed as close to R as possible; there are 3.2^{j-1} nodes at distance j from R for $j = 1,2,3,\ldots,k-1$.

If the kth level contains 3.2^{k-1} nodes, we speak of a complete Moore graph; otherwise, the graph is incomplete. We employ the notation $M(N,3)$ for a generalised Moore graph on N nodes (complete if and only if $N = 3.2^k-2$).

Furthermore, we require that the distance pattern be the same no matter what node is chosen as root node R. This is the essential graph-theoretic condition.

3. COMPLETE MOORE GRAPHS

In a complete Moore graph, the maximal distance from R is k, and the smallest circuit is a $(2k+1)$-gon. If we count all these circuits, their number is easily determined.

Number of nodes in graph = $N = 3.2^k-2$.

Number of edges in graph = $\frac{3N}{2} = 3(3.2^{k-1}-1)$.

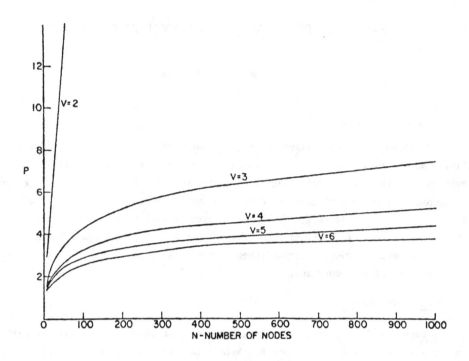

FIGURE 1 - Plot of P(N,V)

Number of edges in joins at levels lower than $k = 3 + 6 + 12 + \ldots + 3.2^{k-1}$
$$= 3(2^k-1).$$

Number of edges at level $k = 3(3.2^{k-1}-2^k) = 3.2^{k-1}$.

Each edge at level k determines one (2k+1)-gon through R. Hence:
Number of (2k+1)-gons through $R = 3.2^{k-1}$.

Total number of (2k+1)-gons $= 3.2^{k-1}(3.2^k-2)$.

Total number of distinct (2k+1)-gons $= \dfrac{3.2^k(3.2^{k-1}-1)}{2k+1}$.

We thus have the

THEOREM. *A complete Moore graph can only exist when* $2k+1$ *divides* (3.2^{k-1}).
We note that this occurs (for $k \leqslant 125,000)$ *only for* $k = 1,2,6425,$ *and* 116983.

The first 2 cases lead to the tetrahedral graph and the Petersen graph. None of the others can exist, since they would be minimal graphs of girth $2k+1$ and valence 3; it is known from various results, summarised in Biggs [1], that graphs satisfying this girth condition can only exist (for $k = 3$) when the girth does not exceed 5.

4. MANUAL RESULTS

Papers [2]-[7] give a thorough discussion of possible graphs on N vertices for $N = 4(2)24$. As a summary of results the number of distinct graphs on N vertices is given as $f(N)$ in the following table.

N	4	6	8	10	12	14	16	18	20	22	24
f(N)	1	2	2	1	2	8	6	1	1	0	1

In addition to these published results, D.D. Cowan has determined that $f(26) = 2$. Also, examples of graphs on 28, 30, 32, 34 nodes have been given; these were obtained in several cases by a type of inductive heuristic approach.

By the time one reaches the case $N = 26$, the case approach has become quite complicated. The main problem is that the "same" graph occurs many times, and so isomorphism rejection is needed again and again.

5. COMPUTERISATION OF THE PROCEDURE

The type of argument used in [2]-[7] can be computerised readily, and employed in conjunction with the fast isomorphism algorithm described by McKay [8]. That algorithm has handled isomorphism of graphs with up to 3000 vertices, and also has the advantage of producing the automorphism group aut (G) of the graph G.

When the graphs were determined by computer, previous results for $N \geq 20$ were verified, and a number of new ones obtained, as summarised in the following table. Here N denotes the number of vertices in G, s is the number of orbits in aut (G), g is the order of aut (G).

N	s	g	N	s	g	N	s	g
4	1	24	14	1	336	16	4	6
6	1	72	14	4	4	16	6	4
6	1	12	14	3	8	16	4	6
8	3	12	14	2	14	16	6	4
8	1	16	14	3	8	16	9	2
10	1	120	14	8	2	18	4	8
12	2	18	14	5	4	20	3	20
12	2	16	14	6	4	24	2	32
			16	9	2			

The new results are as follows.

N	s	g
26	3	48
26	1	52
28	16	2
28	15	2
28	16	2
28	15	2
28	5	12
28	10	4
28	15	2
28	1	336
28	3	14

we thus see that $f(28) = 9$. A subsidiary program was used to produce the largest cycle in the graph. It turns out that the Petersen graph and the Coxeter graph are the only non-Hamiltonian graphs for $N \leqslant 30$. This gives the

THEOREM. *For $N \leqslant 30$, all graphs* $M(N,3)$ *are Hamiltonian save the Petersen graph and the Coxeter graph.*

6. THE CASE N = 30

The case $N = 30$ is particularly interesting since there are 40 distinct generalised Moore graphs with aut (g) ranging from 1440, in the case of the 8-cage, down to the identity. The N, s, g table follows (N is omitted, since $N = 30$ in all cases).

s	g	s	g	s	g	s	g
1	1440	16	2	16	2	30	1
5	16	16	2	30	1	15	2
4	16	10	4	30	1	8	6
9	4	17	2	15	2	12	3
5	8	30	1	18	2	30	1
4	16	16	2	30	1	17	2
15	2	10	4	30	1	15	2
9	4	16	2	30	1	17	2
18	2	16	2	17	2	17	2
10	4	16	2	5	8	30	1

The case $N = 30$ is perhaps of sufficient complexity to merit a further table. The algorithm looks at the possible top layers; for each top layer, A generalised Moore graphs are determined, B of these being non-isomorphic; the time employed is given as t. The total for the B column is 355, but each graph may occur in several rows, and only 40 of the graphs are non-isomorphic. The total time is 13,238 seconds (about 3.7 hours); this time was on a CDC Cyber 73. For running times on the 370/168, divide by 5, roughly.

Possible top layers	A	B	t
1) o o o o o o o o o	2	1	1
2) o o o o o o o o—o	6	1	3
3) o o o o o o o—o—o	79	2	96
4) o o o o o o—o o—o	123	5	194
5) o o o o o o—o—o—o	59	6	199
6) o o o o—o o—o—o	59	10	413
7) o o o o o—o—o—o—o	14	9	228
8) o o o—o o—o o—o	139	17	1207
9) o o o—o—o o—o—o	199	18	991
10) o o o—o o—o—o—o	281	32	1807
11) o o o—o—o—o—o—o	81	21	824
12) o o—o o—o o—o	163	28	1199
13) o o—o o—o—o—o—o	70	26	732
14) o o—o—o o—o—o—o	134	31	566
15) o o—o—o—o—o—o—o	52	16	754
16) o o—o—o—o—o—o	2	1	89
17) o—o o—o o—o o—o	219	15	561
18) o—o o—o—o o—o—o	83	16	582
19) o—o o—o o—o—o—o	124	20	920
20) o—o o—o—o—o—o—o	107	25	748
21) o—o—o o—o—o—o—o	25	14	263
22) o—o—o—o o—o—o—o	153	26	534
23) o—o—o—o—o—o—o—o	34	11	303
24) o—o—o—o—o—o—o	7	4	24
	2215	355	13238
		(40)	

All Hamiltonian

TABLE FOR N = 30

7. CASES WITH N > 30

For N = 32, we hope to obtain results by rewriting the algorithms to speed up the search. However, there are some interesting results in the neighbourhood of N = 46 (no graph) which differ from the results in the neighbourhood of N = 22. We first prove

LEMMA 1. *In the case of* $M(3.2^k-4,3)$, *we may always assume that there are no joins at level* $k-1$, *provided that* $3 < k < 6426$.

Proof. Let there be a joins at level k, c at level k - 1, b between levels k and k - 1. Then

$$2a + b = 3(3.2^{k-1}-2),$$
$$2c + b = 3.2^{k-1}.$$

However, $b \geqslant 3.2^{k-1} - 2$; hence we have two possibilities, namely,

$$b = 3.2^{k-1} - 2, \quad c = 1, \quad a = 3.2^{k-1} - 2;$$
$$b = 3.2^{k-1}, \quad c = 0, \quad a = 3.2^{k-1} - 3.$$

Suppose, if possible, that we always have the first case, no matter what our choice of root node R. Then we can count the number of (2k-1)-gons in the graph; it is simply

$$\frac{3.2^k - 4}{2k - 1},$$

since we get one such circuit for each choice of root node and each circuit is counted 2k - 1 times. Thus it is clear that 2k - 1 must divide $3.2^{k-2} - 1$; for k = 2,3, this does occur (cf. $M(8,3), M(20,3)$). This cannot occur for $3 < k < 6426$.

Thus, we may assume that there exists a vertex R such that there are no joins at level k - 1.

We have not completed a discussion of $M(44,3)$, but this lemma shows that we may consider the case of a root node R for which the 22 vertices at level 4 have 21 lines at level 4, 24 lines joining level-4 vertices to level-3 vertices.

We now pass to the casee of 3.2^k vertices, that is the case just above a complete Moore graph. We restrict ourselves to $m(48,3)$, that is k = 4.

First, we deal with the case a = 1, b = 4, c = 22. Then there are 2 distinct ways of joining the vertices at level r.

Case A

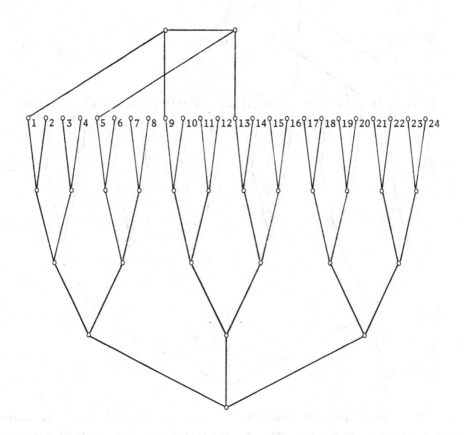

Vertex 1 must join to {17,...,24}; so join 1-17, be equivalence.

Then 9 must join to {21,...,24}; so join 9 to 21. Vertex 17 can then go to either 15 or 16; so join 17 to 15; similarly, 21 can go to either 7 or 8; so join 21 to 7. Now 15 is forced to join to 8, and 7 cannot be joined to any vertex.

Case B.

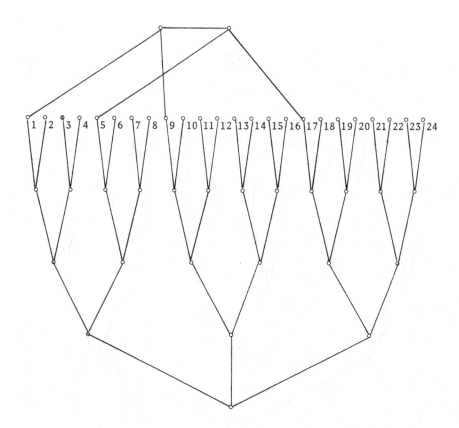

In this case, the subtrees containing {1,...,8}, {9,...,16}, {17,...,24}, are referred to as the *flowers* F_1, F_2, F_3, respectively. Any vertex not joined to level 5 must be adjacent to a vertex of each of the flowers not containing it.

As in Case A, we find (using equivalence) a forced sequence of edges. We join 9 to 21. Then, with no loss of generality, 10 is joined to 19 in F_3. We can then join 21 to 7, 17 to 13, 13 to 3, 1 to 15, 5 to 23, 10 to 4. Again, without loss of generality, 18 may be joined to F_2 at 11; then we join 18 to 8, 3 to 22; and 4 can not be joined to any vertex.

An alternative forced sequence is 3-17; 4-21; 2-19; 1-23; 23-15; 24-11; 15-7; 24-6. Then 6 has no possible join.

Now suppose that, for all R, all the 6 lines from A and B, the two vertices at level 5, go to distinct level-4 vertices. This leaves 18 vertices at level 4 which have two level-4 edges emanating from them. We count decagons.

Decagons through R either go to A (three) or B (three) or have two edges at level 4 (eighteen, one for each of the level-4 vertices which has two level-4 lines from it). Thus the total number of decagons is 24 (48)/10, and this is impossible. Thus we have

LEMMA 3. *There is a root R for which (exactly) one level-4 vertex is joined to both A and B.*

We now discuss this case, illustrated below.

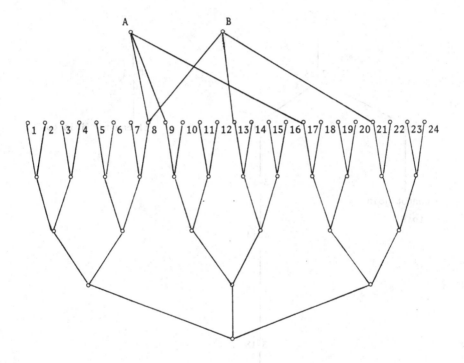

We represent forced joins by a tree (using equivalence).

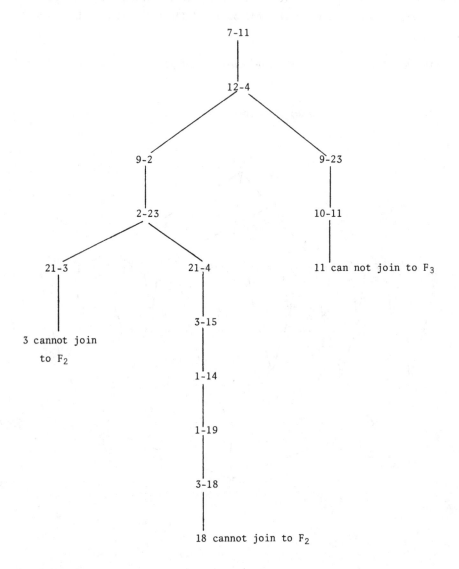

We thus have the

THEOREM.   *A generalised Moore graph* M(48,3) *on 48 vertices does not exist.*

The result of this theorem has been verified by running a programme analogous to that used to produce results on 30 vertices;   the same programme also established the

THEOREM.   *The generalised Moore graph* M(50,3) *does not exist.*

REFERENCES

[1]   N.L. Biggs, Algebraic Graph Theory, Cambridge Tracts in Mathematics No. 67 (Cambridge University Press, London, 1974).

[2]   V.G. Cerf, D.D. Cowan, R.C. Mullin, R.G. Stanton, Computer Networks and Generalised Moore Graphs, *Proceedings Third Manitoba Conference on Numerical Mathematics, Congressus Numerantium IX* (Winnipeg, 1973), 379-398.

[3]   V.G. Cerf, D.D. Cowan, R.C. Mullin, R.G. Stanton, Topological Design Considerations in Computer Communications Networks, *Computer Communication Networks* (ed. R.L. Grimsdale and F.F. Kuo), Nato Advanced Study Institute Noordhoff, Leyden (1975).

[4]   V.G. Cerf, D.D. Cowan, R.C. Mullin, R.G. Stanton, *A Partial Census of Trivalent Generalized Moore Graphs, Invited Address, Proc. 3rd Australian Combinatorial Conference* (Brisbane, 1974), (Springer-Verlag), 1-27.

[5]   V.G. Cerf, D.D. Cowan, R.C. Mullin, R.G. Stanton, Trivalent Generalised Moore Networks on 16 Nodes, *Utilitas Math.* 6 (1974).

[6]   V.G. Cerf, D.D. Cowan, R.C. Mullin, R.G. Stanton, A Lower Bound on the Average Shortest Path Length in Regular Graphs, *Networks*, Vol. 4, No. 4 (1974), John Wiley & Sons, 335-342.

[7]   V.G. Cerf, D.D. Cowan, R.C. Mullin, R.G. Stanton, Some Extremal Graphs, *Ars Combinatoria* 1 (1976), 119-157.

[8]   B.D. McKay, Computing Automorphisms and Canonical Labelling of Graphs, to appear, *Proc. International Conference on Combinatorics* (Canberra, 19777).

SOME 2-(2N+1,N,N-1) DESIGNS WITH MULTIPLE EXTENSIONS

D.R. BREACH

     A 2-(2n+1,n,λ) design can always be extended to a 3-(2n+2,n+1,λ) design by complementation. If λ is large enough there may be other methods of extension. By constructing non-self-complementary 3-(18,9,7) designs it is shown that there is a 2-(17,8,7) design with 16 extensions. The method generalises to give non-self-complementary 3-(2n+2,n+1,n-1) designs for larger values of n.

     A t-(v,k,λ) design on v varieties (or points or symbols) consists of k-sets, called blocks, chosen from the v variaties in such a way that all the blocks are different and each t-subset of varieties occurs exactly λ times in the design. In this paper structures resembling t-designs but not having all the required properties will be called arrays. The meaning of the word "array" each time it is used should be clear from its immediate context.

     In a t-design let λ_i $(0 \leq i \leq t)$ be the number of times each unordered i-subset occurs; in particular, $\lambda_0 = b$, the number of blocks, and $\lambda_1 = r$, the number of replications of each variety. Then the λ_i's are given by the standard equations

$$\lambda_i = \frac{(v-i)(v-i-1)(v-i-2)\ldots(v-t+1)}{(k-i)(k-i-1)(k-i-2)\ldots(k-t+1)}\lambda$$

with, of course, $\lambda_t = \lambda$. If the design contains all possible k-sets from v varieties then it is said to be trivial.

     From a given t-design D a (t-1)-design can be constructed by discarding all blocks not containing a given variety x which is then deleted from the blocks that remain. The design so obtained is called a restriction (or contraction) of D on x and is denoted by D_x. The reverse process can sometimes be performed by adding a new variety x to all the blocks in a (t-1)-design to form new blocks which are supplemented by further new blocks not containing x to make a t-design. In this case the new design is called an extension of the original design.

     In an extension by complementation each original block has the same new variety added to it. Then complements with respect to the extended variety set are taken to form further new blocks. The resulting design is self-complementary and each of its blocks contains half the total number of varieties.

     For 2-$(2n+1,n,\lambda)$ designs there are two well-known results about extensions to 3-designs. They are:

     (i)   any 2-$(2n+1,n,\lambda)$ design can be extended to a 3-$(2n+2,n+1,\lambda)$ design by complementation (Sprott [5]);

     (ii) for $2\lambda = n-1$ (i.e., for Hadamard designs) there is only one way of extending to a 3-design and that is by complementation (Dembowski [2]).

The proofs of these are worth reproducing here since not only are they brief but they also illustrate basic counting principles. To prove (i), let N_3 be the number of blocks in the 2-$(2n+1,n,\lambda)$ design which contain all three of a given triple of varieties. Let N_0 be the number of blocks containing none of them. Then, by the principle of inclusion and exclusion,

$$N_0 = \binom{3}{0} b - \binom{3}{1} r + \binom{3}{2}\lambda - \binom{3}{3} N_3$$

which yields $N_0 + N_3 = \lambda$. Therefore, in the array obtained by complementation there are λ complementary pairs of blocks with one block of each pair containing a given triple and so the array is a 3-design.

The proof of (ii) uses a property of symmetric 2-designs (the designs for which b = v) namely, any block in such a design intersects any other in exactly λ varieties. Now in a 3-$(2n+2,n+1,\tfrac{1}{2}n-\tfrac{1}{2})$ design if two blocks A and B have a variety in common then a restriction on that variety leads to a symmetric 2-design in which any pair of blocks have $(\tfrac{1}{2}n-\tfrac{1}{2})$ varieties in common. Therefore, in the 3-design A and B either intersect in $\tfrac{1}{2}n+\tfrac{1}{2}$ varieties or are disjoint in which case they are complementary. Now take a block A and let N be the number of blocks intersecting it. Then count the pairs (x,Y) where the variety x belongs to both the block Y and the block A with Y $\neq$ A. Then counting these pairs in two ways one has

$$\tfrac{1}{2}N(n+1) = (n+1)(r-1)$$

where r is the replication number for the 3-design and is equal to $(2n+1)$. Thus $N = 4n$. But the total number of blocks b is $(4n+2)$. Of these one is A and 4n are blocks intersecting A so there must be one block not intersecting A. Hence a 3-$(2n+2,n+1,\tfrac{1}{2}n-\tfrac{1}{2})$ design is always self-complementary.

Among the 2-$(2n+1,n,\lambda)$ designs the Hadamard designs are those with the smallest value of λ and their extensions to 3-designs are uniquely determined. For larger values of λ however it may be possible to extend to 3-designs by methods other than by complementation in which case non-self-complementary 3-$(2n+2,n+1,\lambda)$ designs will be formed. Indeed, among the eleven non-isomorphic 2-$(9,4,3)$ designs (Stanton, Mullin and Bate [6] : and also [1] and [4]) there are just two that can be extended in more than one way [1]. One of these has three different extensions, two of which are isomorphic. The other one, which is the genesis of this paper, can be extended in just two ways. This is the baby of a whole family of designs with multiple extensions to 3-designs. The non-self-complementary 3-$(10,5,3)$ design that it generates can be presented by using two sets of five symbols which here will be distinguished by using two differing typefaces 01234 and *01234*. (The reader may find it worthwhile to write the designs in this paper with two different colours of ink, say red and blue.) The design has 36 blocks subdivided into sets of 20 and 16 labelled DD* and H respectively (fig. 1)

A restriction on *0* say, produces a 2-$(9,4,3)$ design which can then be extended by complementation to make a self-complementary 3-$(10,5,3)$ design. As the DD* blocks can be arranged in complementary pairs this design can be formed from Figure 1

by replacing the last 8 blocks in the H section with those in figure 2.   Thus we have
a 2-(9,4,3) design with two different extensions.

0 1 1 4 4	0 2 2 3 3	0 1 2 3 4
0 1 1 4 4	0 2 2 3 3	0 1 2 3 4
1 2 2 0 0	1 3 3 4 4	0 1 2 3 4
1 2 2 0 0	1 3 3 4 4	0 1 2 3 4
2 3 3 1 1	2 4 4 0 0	0 1 2 3 4
2 3 3 1 1	2 4 4 0 0	0 1 2 3 4
3 4 4 2 2	3 0 0 1 1	0 1 2 3 4
3 4 4 2 2	3 0 0 1 1	0 1 2 3 4
4 0 0 3 3	4 1 1 2 2	0 1 2 3 4
4 0 0 3 3	4 1 1 2 2	0 1 2 3 4
	DD* blocks	0 1 2 3 4
		0 1 2 3 4
		0 1 2 3 4
		0 1 2 3 4
		0 1 2 3 4
		0 1 2 3 4
		H blocks

Figure 1   A non-self-complementary 3-(10,5,3) design

$t = 3$, $v = 10$, $k = 5$, $b = 36$, $r = 18$, $\lambda_2 = 8$, $\lambda_3 = 3$

    In figure 1 the DD* blocks contain two sorts of triples, those like xyz and
xyz in which there are 3 distinct numerical symbols and those like xxy in which a
number is repeated.   Triples like xyz, xyz, xyz etc. occur just once in the DD* blocks.
Triples with a repeated number occur 3 times in the DD* blocks.   Therefore, since each
triple occurs 3 times in the whole design the H blocks cannot contain any pairs xx.
Thus each H block contains all five numbers and it is the pattern of type-faces that
really matters for these blocks.   If the type-faces are replaced by +1 and -1 then the
resulting columns (reading down the blocks) are like those that occur in a standardised
Hadamard matrix;   hence, the name H blocks.

    Now for the DD* blocks the permutation (xx) is an automorphism so there is no
loss of generality in taking the first H block to contain symbols of all one type.

0 1 2 3 4	0 1 2 3 4
0 1 2 3 4	0 1 2 3 4
0 1 2 3 4	0 1 2 3 4
0 1 2 3 4	0 1 2 3 4

Figure 2   Replace the last 8 blocks of figure 1 by these blocks to make a self-
complementary 3-(10,5,3) design

The intention is to form the H blocks to complete a non-self-complementary 3-(10,5,3) design. It can be shown [1] that if two blocks of a 3-(10,5,3) design intersect in four varieties then their complements are both in the design. With this result as a guide it is not difficult to complete the H blocks of figure 1.

If in the DD* blocks of figure 1 all the symbols of one type are omitted then the resulting blocks can be rearranged as in figure 3 to form a 2-(5,2,1) design D and its complement a 2-(5,3,3) design D*. The design D although trivial has the property that its blocks can be arranged in disjoint pairs so there is a symbol missing from each pair and each symbol is omitted just once from a disjoint pair. Such a design will be called a block pair disjoint design. The complement D* of such a design is formed by adding the missing symbol of each disjoint pair to each block of the pair.

$$
\begin{array}{cc}
1\ 4 & 2\ 3 \\
2\ 0 & 3\ 4 \\
3\ 1 & 4\ 0 \qquad \text{D} : 2\text{-}(5,2,1) \\
4\ 2 & 0\ 1 \\
0\ 3 & 1\ 2 \\
\end{array}
$$

$$
\begin{array}{cc}
0\ 1\ 4 & 0\ 2\ 3 \\
1\ 2\ 0 & 1\ 3\ 4 \\
2\ 3\ 1 & 2\ 4\ 0 \qquad \text{D*} : 2\text{-}(5,3,3) \\
3\ 4\ 2 & 3\ 0\ 1 \\
4\ 0\ 3 & 4\ 1\ 2 \\
\end{array}
$$

Figure 3   A block pair disjoint design D and its complement D* (with flag-poles to the fore).

In the blocks of D* these symbols occupy special positions which will be called flag-poles. Thus each symbol is a flag-pole twice. The DD* blocks of figure 1 are constructed by a duplication of the numbers of the D* blocks which are not flag-poles followed by an interchange of type-faces to produce further blocks.

Note that, although the circumstances are trivial, the array of figure 3 is a 3-array and would be a 3-design formed by complementation if an extra symbol were added to the blocks of D.

These details of structure will now be mimicked to form 3-(2n+2,n+1,n-1) designs which are not self-complementary and whose existence implies the existence of 2-(2n+1,n,n-1) designs with multiple extensions. The method will be demonstrated by constructing non-self-complementary 3-(18,9,7) designs from which we derive a 2-(17,8,7) design with 16 different extensions.

For a 3-(18,9,7) design the parameters are v = 18, b = 68, k = 9, r = 34, λ_2 = 16, λ_3 = 7. We start by constructing 36 DD* blocks and then to these add different sets of H blocks. A block pair disjoint 2-(9,4,3) D is needed. There is just one such design up to an isomorphism (see figure 4).

Flag-poles are added to the blocks to form the complementary design D* which is a 2-(9,5,5). Then in the blocks of D* the numbers which are not flag-poles are duplicated in the other type-face and finally the type-faces are interchanged to make further blocks. Thus the 36 DD* blocks of figure 5 are formed.

2 3 4 5	1 6 7 8	D 2-(9,4,3)
5 8 2 7	0 6 3 4	
0 3 1 7	4 8 5 6	v = 9
0 2 6 8	5 7 1 4	b = 18
6 7 0 5	2 8 1 3	r = 8
1 8 0 4	3 7 2 6	k = 4
4 7 3 8	0 1 2 5	λ = 3
4 6 1 2	3 5 0 8	
1 5 3 6	2 4 0 7	
0 2 3 4 5	0 1 6 7 8	D* 2-(9,5,5)
1 5 8 2 7	1 0 6 3 4	
2 0 3 1 7	2 4 8 5 6	v* = 9
3 0 2 6 8	3 5 7 1 4	b* = 18
4 6 7 0 5	4 2 8 1 3	r* = 10
5 1 8 0 4	5 3 7 2 6	k* = 5
6 4 7 3 8	6 0 1 2 5	λ* = 5
7 4 6 1 2	7 3 5 0 8	
8 1 5 3 6	8 2 4 0 7	

Figure 4   The unique block pair disjoint 2-(9,4,3) design and its complement forming a 3-array.

It is time to pause and count triples. Deleting a type-face from figure 5 produces the 3-array of figure 4 which but for a missing symbol would be a 3-design obtained as an extension by complementation of a 2-(9,4,3) design. Therefore, triples xyz occur 3 times in the DD* blocks. Since $(\bar{x}x)$ is an automorphism for these blocks all triples containing three different numbers appear 3 times. The only other triples are those containing a repeated number, e.g., $\bar{x}xy$. These occur as often as the pair xy provided the x is not a flag-pole. From figure 4 the pair xy occurs 8 times but in just one of these x is a flag-pole. Therefore in the DD* blocks triples $\bar{x}xy$ occur 7 times.   Since for the whole design $\lambda_3 = 7$ the pair $\bar{x}x$ cannot occur again and the remaining 32 blocks must be H blocks.

These 32 blocks fall into two classes of 16 according as they contain 0 or $\bar{0}$. Call these classes H(0) and H($\bar{0}$). No block in either class can contain a repeated number. Such triples that occur in the H blocks must do so 4 times. The trick is to make sure each such triple occurs twice in each of H(0) and H($\bar{0}$) with the proviso that triples containing 0 appear four times in H(0) and similarly for those containing $\bar{0}$ which are in H($\bar{0}$).

Since $(\bar{x}x)$ is an automorphism for the DD* blocks the first H(0) block can be taken to be 012345678.   To keep pairs including 0 balanced between the two type-faces take the last H(0) block to be $0\bar{1}\bar{2}\bar{3}\bar{4}\bar{5}\bar{6}\bar{7}\bar{8}$ (see figure 6). In the remaining 14 blocks

```
0 2 2 3 3 4 4 5 5        0 1 1 6 6 7 7 8 8
0 2 2 3 3 4 4 5 5        0 1 1 6 6 7 7 8 8
1 5 5 8 8 2 2 7 7        1 0 0 6 6 3 3 4 4
1 5 5 8 8 2 2 7 7        1 0 0 6 6 3 3 4 4
2 0 0 3 3 1 1 7 7        2 4 4 8 8 5 5 6 6
2 0 0 3 3 1 1 7 7        2 4 4 8 8 5 5 6 6
3 0 0 2 2 6 6 8 8        3 5 5 7 7 1 1 4 4
3 0 0 2 2 6 6 8 8        3 5 5 7 7 1 1 4 4
4 6 6 7 7 0 0 5 5        4 2 2 8 8 1 1 3 3
4 6 6 7 7 0 0 5 5        4 2 2 8 8 1 1 3 3
5 1 1 8 8 0 0 4 4        5 3 3 7 7 2 2 6 6
5 1 1 8 8 0 0 4 4        5 3 3 7 7 2 2 6 6
6 4 4 7 7 3 3 8 8        6 0 0 1 1 2 2 5 5
6 4 4 7 7 3 3 8 8        6 0 0 1 1 2 2 5 5
7 4 4 6 6 1 1 2 2        7 3 3 5 5 0 0 8 8
7 4 4 6 6 1 1 2 2        7 3 3 5 5 0 0 8 8
8 1 1 5 5 3 3 6 6        8 2 2 4 4 0 0 7 7
8 1 1 5 5 3 3 6 6        8 2 2 4 4 0 0 7 7
```

Figure 5   36 DD* blocks for a 3-(18,9,7) design.

```
0 1 2 3 4 5 6 7 8
0 1 2 3 4 5 6 7 8        1 2 3 4
0 1 2 3 4 5 6 7 8                5 6 7 8
0 1 2 3 4 5 6 7 8        1 2         7 8
0 1 2 3 4 5 6 7 8          3 4 5 6
0 1 2 3 4 5 6 7 8        1   3     6   8
0 1 2 3 4 5 6 7 8          2   4 5   7
0 1 2 3 4 5 6 7 8        1     4   6 7
0 1 2 3 4 5 6 7 8          2 3   5     8
0 1 2 3 4 5 6 7 8        1 2     5 6
0 1 2 3 4 5 6 7 8          3 4     7 8
0 1 2 3 4 5 6 7 8        1   3   5   7
0 1 2 3 4 5 6 7 8          2   4   6   8
0 1 2 3 4 5 6 7 8        1       4 5     8
0 1 2 3 4 5 6 7 8          2 3     6 7
0 1 2 3 4 5 6 7 8
```

H(0) blocks   3-(8,4,1) : $v = 8$, $k = 4$

$b = 14$, $r = 7$

$\lambda_2 = 3$, $\lambda_3 = 1$

Figure 6   A set of H(0) blocks for a 3-(18,9,7) design.

of H(0) triples $0yz$ must occur three times and triples xyz not containing 0 must occur
once. These observations suggest that excluding 0 and concentrating on one type-face

we should have 3-design in these 14 blocks on the symbols *12345678*. But there is a
3-(8,4,1) design which is self-complementary being the extension of a Hadamard
2-(7,3,1) design (see figure 6).

The complement of the 3-(8,4,1) design is the same design so when the 14
blocks are completed with symbols from 12345678 the triple count for that type-face
is correct as well. It remains to check triples containing both type-faces. Any
pair of non-zero numbers appears 16 times (once for each block). Of these, 4 are *xy*,
4 are x*y*, 4 are X*y* and 4 are *xy*. Thus all triples containing *0* appear 4 times in the
H(*0*) blocks. Now take a triple *xyz* not containing *0*. This occurs in two blocks. The
pair *xy* occurs without *z* in 1+3-2 = 2 blocks. Therefore, *xyz* occurs just twice in
the H(*0*) blocks. Thus the H(*0*) blocks have the correct triple count.

To complete the 3-(18,9,7) design a set of 16 H(0) blocks is needed. One way
of making these is to replace *0* by 0 in the H(*0*) blocks (see figure 7(i)). This
produces a self-complementary 3-design. To obtain a non-self-complementary design
interchange the type-faces down the 1-column of the H(0) blocks (see figure 7(ii)).

```
        0 1 2 3 4 5 6 7 8              0 1 2 3 4 5 6 7 8
        0 1 2 3 4 5 6 7 8              0 1 2 3 4 5 6 7 8
        0 1 2 3 4 5 6 7 8              0 1 2 3 4 5 6 7 8
        0 1 2 3 4 5 6 7 8              0 1 2 3 4 5 6 7 8
        0 1 2 3 4 5 6 7 8              0 1 2 3 4 5 6 7 8
        0 1 2 3 4 5 6 7 8              0 1 2 3 4 5 6 7 8
        0 1 2 3 4 5 6 7 8              0 1 2 3 4 5 6 7 8
        0 1 2 3 4 5 6 7 8              0 1 2 3 4 5 6 7 8
        0 1 2 3 4 5 6 7 8              0 1 2 3 4 5 6 7 8
        0 1 2 3 4 5 6 7 8              0 1 2 3 4 5 6 7 8
        0 1 2 3 4 5 6 7 8              0 1 2 3 4 5 6 7 8
        0 1 2 3 4 5 6 7 8              0 1 2 3 4 5 6 7 8
        0 1 2 3 4 5 6 7 8              0 1 2 3 4 5 6 7 8
        0 1 2 3 4 5 6 7 8              0 1 2 3 4 5 6 7 8
        0 1 2 3 4 5 6 7 8              0 1 2 3 4 5 6 7 8
        0 1 2 3 4 5 6 7 8              0 1 2 3 4 5 6 7 8
```

(i) A set of H(0) blocks   (ii) A set of H(0) blocks
to complete to a self-   to complete to a non-
complementary 3-(18,9,7)   self-complementary
design   3-(18,9,7) design

Figure 7   Sample sets of H(0) blocks for making 3-(18,9,7) designs.

In effect this amounts to making H blocks with respect to 0 and the two sets of
symbols 1*2345678* and *1*2345678 so the triple count must be correct.

If a restriction on *0* is taken in these completed 3-designs, then a 2-(17,8,7)
design with at least 2 extensions is obtained. In fact, it has at least 16 extensions
because type-face changes on the columns of the H(0) can be made in many different ways.

However, isomorphic sets of blocks occur under these changes and a careful investigation is needed. To generate the 16 sets of H(0) blocks write out the pattern of type-faces in the H(*0*) blocks as in figure 8. Then interchange the types in the 0-column and any other one or two columns. There are 8 ways of changing one column, other than the 0-column. There are 7 ways of changing two columns since, for example, changing columns 1 and 2 produces the same blocks as does changing columns 3 and 4 or 5 and 6 or 7 and 8. The sets of blocks produced by changing more than two columns (excluding the 0-column) are isomorphic to those produced by changing one or two columns. Thus, changing columns 1, 3 and 5 is equivalent to changing columns 1 and 3 first and then column 5. This is equivalent to changing columns 5 and 7 first and then changing column 5 again. Hence, changing three columns is equivalent to changing one column, and so on. The sixteenth way comes from not changing any column other than the 0-column. Thus a 2-(17,8,7) design with sixteen extensions can be constructed.

```
0 1 2 3 4 5 6 7 8 (column numbers)
1 1 1 1 1 1 1 1 1          Change 0-column and:
1 1 1 1 1 . . . .          No other column    1 way
1 . . . . 1 1 1 1          1 other column     8 ways
1 1 1 . . . . 1 1          2 other columns    7 ways
1 . . 1 1 1 1 . .                             16 ways
1 1 . 1 . . 1 . 1
1 . 1 . 1 1 . 1 .          16 non-isomorphic patterns
1 1 . . 1 . 1 1 .
1 . 1 1 . 1 . . 1
1 1 1 . . 1 1 . .
1 . . 1 1 . . 1 1
1 1 . 1 . 1 . 1 .
1 . 1 . 1 . 1 . 1
1 1 . . 1 1 . . 1
1 . 1 1 . . 1 1 .
1 . . . . . . . .
```

Equivalent column pair changes
1,2 - 3,4 - 5,6 - 7,8
1,3 - 2,4 - 5,7 - 6,8
1,4 - 2,3 - 6,7 - 5,8
1,5 - 2,6 - 3,7 - 4,8
1,6 - 2,5 - 3,8 - 4,7
1,7 - 2,8 - 3,5 - 4,6
1,8 - 2,7 - 3,6 - 4,5

Figure 8   Scheme for generating H(0) blocks from a set of H(*0*) blocks.
The method generalises quite readily. Thus a block pair disjoint 2-(17,8,7)

design exists, and Hadamard matrices of order 16 exist so it is possible to construct 3-(34,17,15) designs which are not self-complementary.

There are infinite families of block pair disjoint 2-(2n+1,n,n-1) designs. For example, if 2n+1 is a prime power put non-zero squares in the corresponding finite field into one block, put the non-zero non-squares into another block and use the algebra of the field to generate more blocks. The details are given, Hall [3], p.200 (in a disguised form, since the context is different). Hadamard matrices and therefore Hadamard designs also exist in infinite families. If it were known that block pair disjoint designs exist for all n or that Hadamard matrices exist for all possible orders then it would be possible to assert that there exist 2-designs with more than any specified number of extensions.

For block pair disjoint 2-(2n+1,n,n-1) designs with 2n+1 ≡ 3 (mod 4) the H blocks contain triples an odd number of times and 3-(4n+2,2n+1,2n-1) designs which are not self-complementary are much harder to construct.

REFERENCES

[1]   D.R. Breach, The 2-(9,4,3) and 3-(10,5,3) designs, *J. Combinatorial Theory* (A) (to appear). (A longer version exists as Research Report CORR 77-11 of the Department of Combinatorics and Optimisation, University of Waterloo, Ontario, Canada.)

[2]   P. Dembowski, *Finite Geometries (Ergebnisse der Mathematik 44)*, Springer Verlag, Berlin, 1968.

[3]   Marshall Hall, Jr., *Combinatorial Theory*, Blaisdell, Waltham, Mass., U.S.A., 1967.

[4]   J.H. van Lint, H.C.A. van Tilborg, and J.R. Wikeman, Block designs with v = 10, k = 5, λ = 4, *J. Combinatorial Theory* (A) 23 (1977), 105-115.

[5]   D.A. Sprott, Balanced incomplete block designs and tactical configurations, *Ann. of Math. Statist.*, 26 (1955), 752-758.

[6]   R.G. Stanton, R.C. Mullin, and J.A. Bate, Isomorphism classes of a set of prime BIBD parameters, *Ars Combinatoria* 2, (1976), 251-264.

Department of Mathematics,
University of Canterbury,
Christchurch,
New Zealand.

05C99

THE GRAPH OF TYPE (0,∞,∞) REALIZATIONS OF A GRAPHIC SEQUENCE

R.B. EGGLETON AND D.A. HOLTON

This paper investigates some of the fundamental relationships between various nonisomorphic pseudographs (generalized graphs in which loops and multiple edges are permitted) which have the same degree sequence.

1. GRAPHIC SEQUENCES OF TYPE (0,∞,∞)

This paper is concerned with pseudographs, that is, generalized graphs in which the only kinds of generalized edge permitted are loops and ordinary edges. We shall use the terminology and notation of our accompanying introductory paper [1], according to which a generalized graph $G = (V,E)$, with edge selection $E \in V^{oo}$, is a pseudograph if it is of type $(0,\infty,\infty)$, that is, every generalized edge in E is similar to $[0^2]$ or $[0,1]$ and multiplicities are unrestricted. As in [1], attention is confined to generalized graphs with locally finite edge selections. In the case of a pseudograph, this means simply that each vertex occurs in only finitely many of the edges.

As first noted by Senior [2], there is a simple characterization of finite sequences which are realizable by graphs of type $(0,\infty,\infty)$. This is covered in the following theorem, which deals with infinite sequences as well. (Throughout this paper, when "infinite" is used without further qualification we shall mean countably infinite.)

THEOREM 1. *The sequence $\underline{d} \in \mathbf{N}^*$ is realizable as a graph of type $(0,\infty,\infty)$ precisely when $\Sigma\underline{d}$ is either even or infinite.*

Proof. If the edge selection E of a locally finite graph of type $(0,\infty,\infty)$ is infinite, its degree sequence $\underline{d}$ has infinite sum. If E is finite, $\underline{d}$ has even sum, since any loop or ordinary edge contributes 2 to the value of $\Sigma\underline{d}$. The stated conditions are therefore necessary.

Now suppose $\underline{d} \in \mathbf{N}^*$ is given, with $\Sigma\underline{d}$ either even or infinite. Let the index set of $\underline{d}$ be V (by convention, an initial segment, or all, of $\mathbf{N}$), and let $V' = \{v_0, v_1, v_2, \ldots\}$ be the subset of V corresponding to the odd terms of $\underline{d}$, that is, d_i is odd precisely when $i \in V'$. Let $\underline{e} \in \mathbf{N}^*$ be such that $e_i = 1$ if $i \in V'$ and $e_i = 0$ otherwise. There are two cases. (1) If $|V'|$ is even or infinite, $\underline{d}$ is realized by $G = (V,E)$, where E comprises the loops $[i^2]^{(d_i-e_i)/2}$ for each $i \in V$, and the edges $[v_{2j}, v_{2j+1}]$, running over the vertices in V'. Evidently G is of type $(0,\infty,\infty)$. (2) If $|V'|$ is odd, $\Sigma\underline{d}$ must be infinite, so there are infinitely many positive even

terms in $\underset{\sim}{d}$. Let v_{2n} be the last (that is, largest) member of V', and let
V" = $\{w_0,w_1,w_2,\ldots\}$ be the subset of V comprising all indices greater than 2n which
correspond to positive even terms of $\underset{\sim}{d}$, that is, d_i is positive and even, with i > 2n,
precisely when $i \in$ V". Then d is realized by G = (V,E), where E comprises the loops
$[i^2]^{(d_i-e_i)/2}$ for $i \in$ V with $i \le 2n$, the loops $[i^2]^{(d_i-2)/2}$ for $i \in$ V", the edges
$[v_{2j},v_{2j+1}]$, running over the vertices in V' which are smaller than 2n, the edge
$[v_{2n},w_0]$, and the edges $[w_k,w_{k+1}]$, running over all the vertices in V". Again G is
of type $(0,\infty,\infty)$. □

The realizations constructed in the proof are illustrated in Figure 1 (where ω
denotes the sequence of natural numbers in order of magnitude). In fact, the constructions in the proof justify the following result.

COROLLARY. *The sequence* $\underset{\sim}{d} \in$ **N*** *is realizable as a graph of type* $(0,\infty,1)$
precisely when $\Sigma\underset{\sim}{d}$ *is either even or infinite.*

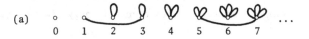

$$(a) \quad \begin{matrix} \circ & & & & & & & & \\ 0 & 1 & 2 & 3 & 4 & 5 & 6 & 7 \end{matrix} \cdots$$

$$(b) \quad \begin{matrix} & & & & & \\ 1 & 2 & 4 & 6 & 8 \end{matrix} \cdots$$

FIGURE 1 - Type $(0,\infty,1)$ realizations of (a) $\underset{\sim}{d} = (0,1,2,\ldots) = \omega$,
(b) $\underset{\sim}{d} = (1,2,4,\ldots) = (1) + 2\omega$.

2. GRAPHS OF TYPE $(0,\infty,\infty)$ REALIZATIONS OF $\underset{\sim}{d}$

If $d \in$ **N*** is realizable as a graph of type $(0,\infty,\infty)$, the study of the relationship of the various realizations of $\underset{\sim}{d}$ is facilitated by considering the simple graph
R = $R(\underset{\sim}{d};(0,\infty,\infty))$, called the *graph of type* $(0,\infty,\infty)$ *realizations of* $\underset{\sim}{d}$. The *vertices*
of R are the nonisomorphic realizations of $\underset{\sim}{d}$, and two realizations G,H are *adjacent*
in R precisely when there exists a *switching* of the form

$$[a,b],[c,d] \rightarrow [a,c],[b,d]$$

which transforms G into H. (The switching selects two generalized edges [a,b], [c,d]
in G, and replaces them by [a,c], [b,d], while leaving the rest of the edge selection
of G unchanged. The vertices a,b,c,d are not necessarily distinct.) There are five
kinds of switching incorporated in this definition. They are shown in Figure 2. All
other possibilities (such as when a = b = c) are invariant under switching. Senior

[2] called these switchings *"transformations"*.

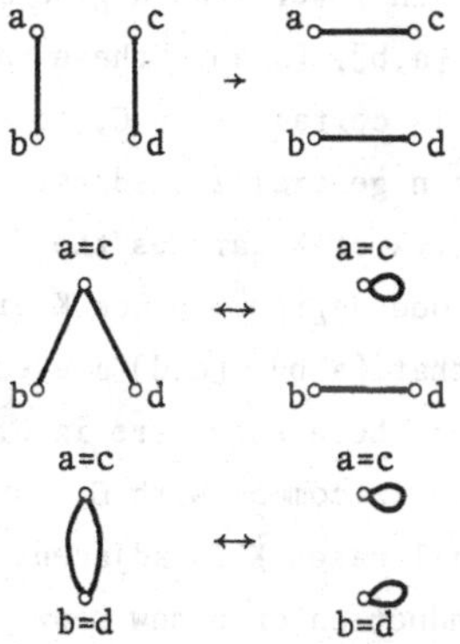

FIGURE 2 - The five kinds of switching (the first is self-inverse, the others not)

3. COMPONENTS OF THE GRAPH $R(\underset{\sim}{d};(0,\infty,\infty))$

Two realizations of a graphic sequence $\underset{\sim}{d}$ are *associates* if they differ at only a finite number of vertices (up to isomorphism). Any class of associates is countable. The operation of switching defined above permits passage from any locally finite pseudograph to any of its associates of the same type (up to isomorphism) in a finite number of steps. In terms of the graph of realizations of a sequence $\underset{\sim}{d}$, this may be stated as follows.

THEOREM 2. *The classes of associates of type* $(0,\infty,\infty)$ *realizations of a sequence* $\underset{\sim}{d} \in \mathbb{N}^*$ *are the components of the graph* $R(\underset{\sim}{d};(0,\infty,\infty))$.

Proof. Vertices of a graph are connected precisely when there is a path of finitely many edges between them. Since a switching alters a pseudograph at only four or fewer vertices, two type $(0,\infty,\infty)$ realizations of $\underset{\sim}{d}$ which are adjacent vertices of R are associates, and hence all vertices connected to a given vertex in R are associates.

Now suppose G,H are type $(0,\infty,\infty)$ realizations of $\underset{\sim}{d}$ which are associates. It may be assumed that G,H differ at only finitely many vertices. We shall show that they are connected vertices of R. If G ∩ H is the subgraph on which G and H coincide, then $G' = G \setminus G \cap H$ and $H' = H \setminus G \cap H$ are type $(0,\infty,\infty)$ realizations of a finite sequence $\underset{\sim}{d}'$ which differ at every vertex. Evidently G,H are connected in $R(\underset{\sim}{d};(0,\infty,\infty))$ if G',H' are connected in $R' = R(\underset{\sim}{d}';(0,\infty,\infty))$. Since $\underset{\sim}{d}'$ is finite, it has an even number of odd terms, so has a realization C as constructed in case (1) of the proof of Theorem 1. Each vertex of C carries the maximum number of loops permitted by its degree, and ordinary edges are disjoint and join consecutive vertices of odd degree. For some $n > 0$, suppose all type $(0,\infty,\infty)$ realizations of $\underset{\sim}{d}'$ which differ from C in fewer than n generalized edges are connected to C in R', and suppose K is a realization which differs from C in precisely n generalized edges. (Note that switching preserves the total number of generalized edges.) If K has a pair of parallel ordinary edges, say $[a,b]^2$, these can be switched to two loops $[a^2]$, $[b^2]$, which are necessarily in C, so the resultant realization K' differs from C in fewer than n generalized edges. If K has a pair of adjacent ordinary edges, say $[a,b]$, $[a,d]$, these can be switched to an ordinary edge $[b,d]$ and a loop $[a^2]$, which is certainly in C, so again the resultant realization K' differs from C in fewer than n generalized edges. If the ordinary edges of K are all disjoint, then every vertex of K carries the maximum number of loops, and ordinary edges join vertices of odd degree. Since K differs from C, there is some earliest odd degree vertex a such that $[a,b]$, $[c,d]$ are edges in K, with c the first odd degree vertex after a. Neither of these edges are in C. Switching to $[a,c]$, $[b,d]$ gives a realization K' which has $[a,c]$ in common with C, so differs from C in fewer than n generalized edges. Thus, in all cases K is adjacent to K', and K' is connected to C, so K is connected to C. Induction on n now shows that all vertices of R' are connected to C, to R' is a connected graph. It follows that H',K' are connected in R', so H,K are connected in R, as required. □

The proof of Theorem 2 shows that the graph $R(\underline{d};(0,\infty,\infty))$ is connected if $\underline{d}$ is finite. More generally, we have

THEOREM 3. *Let $\underline{d}$ be a graphic sequence of type $(0,\infty,\infty)$. Its graph of realizations $R(\underline{d};(0,\infty,\infty))$ is connected precisely when $\underline{d}$ has only finitely many terms greater than 1; otherwise, R has uncountably many components.*

Proof. If $\underline{d}$ has only finitely many terms greater than 1, either $\Sigma\underline{d}$ is finite and so all realizations are associates, or else $\Sigma\underline{d}$ is infinite and has infinitely many terms equal to 1. In the latter case, $\underline{d}$ is equivalent to a finite sequence $\underline{d}'$ followed by an infinite sequence $\underline{d}''$, where $\underline{d}'$ comprises all terms of $\underline{d}$ greater than 1 (with sum s, say) together with s other terms each equal to 1, while $\underline{d}''$ comprises infinitely many terms equal to 1 and possibly has zero terms as well. Clearly any realization of $\underline{d}$ is isomorphic to a realization of $\underline{d}'$ together with a realization of $\underline{d}''$. But all realizations of $\underline{d}''$ are isomorphic, and $\underline{d}'$ is finite, so again all realizations of $\underline{d}$ are associates. Theorem 2 ensures in each case that R is connected.

Now suppose $\underline{d}$ has infinitely many terms greater than 1. Then $\underline{d}$ can be decomposed into two infinite sequences $\underline{d}'$, $\underline{d}''$ such that all terms of $\underline{d}'$ are greater than 1 and have the same parity, while $\underline{d}''$ has infinite sum. Then $\underline{d}'$, $\underline{d}''$ both have type $(0,\infty,\infty)$ realizations. Suppose we can find uncountably many realizations of $\underline{d}'$ no two of which are associates. Then by combining a fixed realization of $\underline{d}''$ with each, we obtain representatives of uncountably many associate classes of realizations of $\underline{d}$, each a vertex in a distinct component of R (by Theorem 2).

It remains to find the desired realizations of $\underline{d}'$. First suppose the terms in $\underline{d}'$ are even. Let S be any nonempty subset of positive integers. For each n ϵ S we can take any n terms of $\underline{d}'$ and realize them by an n-cycle with the appropriate numbers of loops on its vertices (see Figure 3). Thus we can realize $\underline{d}'$ so that for each $n \in S$ it has infinitely many components which are n-cycles (with loops if necessary), and every component is of this form.

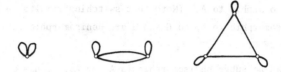

FIGURE 3 - Examples of n-cycles (with all degrees 4) when n = 1,2,3

Realizations of $\underline{d}'$ which correspond to different choices of S differ in infinitely many of their components, so are not associates, and there are uncountably many choices of S. A similar construction can be used if the terms in $\underline{d}'$ are odd, but in this case for each $n \in S$ we take any 2n terms of $\underline{d}'$ and realize them by a pair of n-cycles (with loops if necessary), with the vertices of one joined in some order to those of the other (see Figure 4). □

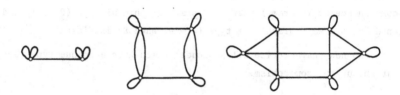

FIGURE 4 - Examples of paired n-cycles (with all degrees 5) when n = 1,2,3

4. UNIQUE TYPE $(0,\infty,\infty)$ REALIZABILITY

If $\underline{d}$ has a unique type $(0,\infty,\infty)$ realization, up to isomorphism, then $R(\underline{d};(0,\infty,\infty))$ is the trivial graph, that is, a single vertex.

Consider the four graphs in Figure 5. They cannot occur as subgraphs in a unique type $(0,\infty,\infty)$ realization, and this will enable us to characterize sequences which have a unique realization.

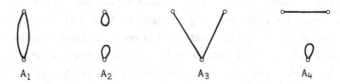

$\qquad A_1 \qquad\qquad A_2 \qquad\qquad A_3 \qquad\qquad A_4$

FIGURE 5 - Four forbidden subgraphs for unique type $(0,\infty,\infty)$ realizability

THEOREM 4. *If G is the unique type* $(0,\infty,\infty)$ *realization of the graphic sequence* $\underset{\sim}{d}$, *up to isomorphism, then G has no subgraph isomorphic to* A_1, A_2, A_3 *or* A_4.

Proof. On the contrary, suppose G has a subgraph A isomorphic to A_1 or A_2. Let H result from switching every member of a maximal edge-disjoint family of subgraphs of G isomorphic to A. (Note that switching results in $A_1 \leftrightarrow A_2$.) Then H has no subgraph isomorphic to A, so G and H are nonisomorphic realizations of $\underset{\sim}{d}$, contrary to the choice of G.

If G has no subgraph isomorphic to A_1 or A_2, it has at most one vertex which carries loops. Suppose G has a subgraph A isomorphic to A_3 or A_4. If H results from switching A, then G and H have different numbers of loops, since switching results in $A_3 \leftrightarrow A_4$. Thus G and H are nonisomorphic realizations of $\underset{\sim}{d}$, again contrary to the choice of G. $\square$

Note that in this proof it is in general not sufficient to switch a finite number of subgraphs isomorphic to A_1 or A_2, since the resultant realization could still be isomorphic to the original if there are infinitely many such subgraphs present.

Given two sequences $\underset{\sim}{d}, \underset{\sim}{e} \in \mathbb{N}^*$, we say that $\underset{\sim}{d}$ *dominates* $\underset{\sim}{e}$ if every pair of corresponding terms satisfies $d_i \geq e_i$ (and if the sequences have different lengths, each unmatched term in the longer is treated as though it is matched with zero). It is an easy consequence of Theorem 1 that if $\underset{\sim}{d}, \underset{\sim}{e}$ are graphic of type $(0,\infty,\infty)$, and $\underset{\sim}{d}$ dominates $\underset{\sim}{e}$, then $\underset{\sim}{d} - \underset{\sim}{e}$ is always graphic of type $(0,\infty,\infty)$ when $\Sigma \underset{\sim}{e}$ is finite.

We can now characterize all sequences which have a unique type $(0,\infty,\infty)$ realization, up to isomorphism.

THEOREM 5. *If* $\underset{\sim}{d}$ *is a graphic sequence of type* $(0,\infty,\infty)$, *then* $R(\underset{\sim}{d};(0,\infty,\infty))$ *is the trivial graph precisely when* $\underset{\sim}{d}$ *has no subsequence equivalent to one which dominates* $(2,1,1)$ *or* $(2,2)$.

Proof. The necessity of the stated conditions follows from Theorems 1 and 4 in view of the fact that A_1, A_2, A_3, A_4 are the only type $(0,\infty,\infty)$ realizations of the sequences $(2,1,1)$ and $(2,2)$. If $\underset{\sim}{d}$ has a subsequence equivalent to one which dominates $(2,1,1)$ or $(2,2)$, it has a realization which has a subgraph A isomorphic to one of the four specified graphs.

The sufficiency of the stated conditions follows from examination of the sequences not excluded by them. If $\underset{\sim}{d}$ is such a sequence with at least three positive terms then every positive term of $\underset{\sim}{d}$ is equal to 1, and all realizations of $\underset{\sim}{d}$ are clearly isomorphic. If $\underset{\sim}{d}$ has precisely two positive terms, one of these terms must be equal to 1 and the other must be odd. If $\underset{\sim}{d}$ has precisely one positive term, this term must be even. The only remaining possibility is that every term of $\underset{\sim}{d}$ is equal to 0. In these three latter cases $\underset{\sim}{d}$ admits no switching, so its type $(0,\infty,\infty)$ realization is unique (see Figure 6). $\square$

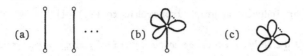

FIGURE 6 - Unique type (0,∞,∞) realizations (without isolated vertices)

5. ISOLATED VERTICES IN R(d̰;(0,∞,∞))

Throughout this section we are concerned with the case in which R(d̰;(0,∞,∞)) has more than one component, so d̰ has infinitely many terms greater than 1 (by Theorem 3).

A type (0,∞,∞) graph G is *invariant under switching* if any graph which results from it by any possible switching is isomorphic to G. The graphs in Figure 6 are invariant under switching, but they are certainly not the only ones. It is possible for a sequence d̰ to have a type (0,∞,∞) realization which is invariant under switching, yet which is not a unique realization of d̰.

An important example of this will now be described. Let d̰ be a sequence in which each positive term is equal to infinitely many other terms. Let F(d̰) be the type (0,∞,∞) realization of d̰ comprising the union of infinitely many disjoint copies of all nonisomorphic type (0,∞,∞) realizations of each finite subsequence of positive terms of d̰ with even sum, together with an isolated vertex for each zero term of d̰. We call F(d̰) the *focus* of d̰.

THEOREM 6.   *If d̰ ∈ N* is a sequence in which each positive term is equal to infinitely many other terms, the focus F(d̰) is a type (0,∞,∞) realization of d̰ which is invariant under switching. Any other such realization of d̰ has F(d̰) as an induced subgraph.*

Proof.   It is clear that F(d̰) is invariant under switching. To see that it is an induced subgraph of any type (0,∞,∞) realization G of d̰ which is invariant under switching, note first that if d̰ has any positive even terms then any particular even degree vertex in G can be isolated by a finite sequence of switchings (after which it carries an appropriate number of loops), so G must already contain infinitely many isolated vertices of each positive even degree present in d̰, since G is invariant under switching. Similarly, if d̰ has any odd terms, any particular pair of odd degree vertices in G can be moved into a single component by a finite sequence of switchings, so G already contains infinitely many components comprising a pair of vertices of any odd degrees present in d̰. Combining these components, G has an induced subgraph which is a realization of any finite subsequence of positive terms of d̰ with even sum, so G must contain infinitely many disjoint copies of each such realization and each of its associates, again since G is invariant under switching. It follows that G has an induced subgraph isomorphic to $F(\underset{\sim}{d})$. $\square$

THEOREM 7. *Let* $\underset{\sim}{d} \in \mathbb{N}^*$ *be a sequence with infinitely many terms greater than 1. Then its graph of realizations* $R(\underset{\sim}{d}; (0,\infty,\infty))$ *has an isolated vertex precisely when (i) each positive term of* $\underset{\sim}{d}$ *is equal to infinitely many other terms of* $\underset{\sim}{d}$, *or else (ii)* $\underset{\sim}{d}$ *has exactly one term equal to 1 and all other positive terms equal to 2.*

Proof. If each positive term of $\underset{\sim}{d}$ is equal to infinitely many other terms, then the focus $F(\underset{\sim}{d})$ is an isolated vertex of R.

Now suppose some positive term of $\underset{\sim}{d}$, say d_n, is equal to only finitely many other terms of $\underset{\sim}{d}$. (1) If $d_n > 1$, switching can always occur on a corresponding vertex in any realization of $\underset{\sim}{d}$, to alter the number of loops it carries; since there are only finitely many vertices of the degree in question, the resulting realization is not isomorphic to the original, so neither is an isolated vertex in R. (2) If $d_n = 1$ and more than one such term occurs in $\underset{\sim}{d}$, any realization can be switched to one in which the number of adjacent pairs of vertices of degree 1 is altered, so no realization is an isolated vertex in R. (3) Finally, suppose $d_n = 1$ and all other positive terms of $\underset{\sim}{d}$ are greater than 1. If $\underset{\sim}{d}$ has any term greater than 2, then any type $(0,\infty,\infty)$ realization G can be switched (if necessary) so that the vertex of degree 1 is adjacent to a vertex of degree greater than 2, and further switching can preserve this adjacency while altering the number of loops carried by the higher degree vertex, so G is not an isolated vertex of R. The remaining possibility is that $\underset{\sim}{d}$ has no term greater than 2, so has infinitely many terms equal to 2. In this case any type $(0,\infty,\infty)$ realization must have the vertex of degree 1 as endvertex of a one-way infinite path P'_∞ (Figure 7). Let H be the realization of $\underset{\sim}{d}$ which comprises P'_∞, together with the focus $F(\underset{\sim}{2})$, where $\underset{\sim}{2} = (2,2,2,\ldots)$. It can be verified that H is invariant under switching, so is an isolated vertex of R. This completes the examination of all possibilities for $\underset{\sim}{d}$. $\square$

This proof brings to our attention the *one-way infinite path* $P'_\infty = (\mathbb{N},E')$, where E' comprises $[i,i+1]$ for each $i \in \mathbb{N}$. In order to count the isolated vertices of R, we shall use constructions involving some related graphs, which we now define. (In some of these definitions it is convenient to use $\mathbb{Z}$ as vertex set rather than $\mathbb{N}$, though isomorphic graphs with $\mathbb{N}$ as vertex set could obviously be specified.)

Let $P_\infty = (\mathbb{Z},E)$, where E comprises $[i,i+1]$ for each $i \in \mathbb{Z}$. This is the *two-way infinite path* (Figure 7).

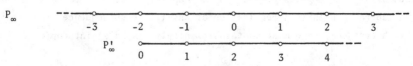

FIGURE 7 - The two-way and one-way infinite paths

For any positive integers a,n, with a > 2, let $H_n(a) = (\mathbb{Z},E)$, where E comprises [i,i+1] for each $i \in \mathbb{Z}$, together with $[2jn+k,2(j+1)n-k-1]^{a-2}$ for each $j,k \in \mathbb{Z}$ with $0 \leqslant k < n$. Note that the multiplicity of each $[(2j+1)n-1,(2j+1)n]$ is equal to a - 1 as a result of this specification of E. Similarly, let $H_n'(a) = (\mathbb{N},E')$ where E' comprises just those edges in E which have both vertices in $\mathbb{N}$ (Figure 8).

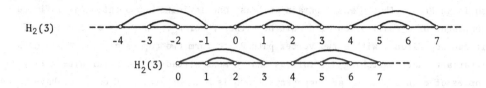

FIGURE 8 - The graphs $H_2(3)$ and $H_2'(3)$

THEOREM 8. *Let $\underset{\sim}{d} \in \mathbb{N}^*$ be a sequence with infinitely many terms greater than 1, and graph of realizations $R(\underset{\sim}{d};(0,\infty,\infty))$. (i) If $\underset{\sim}{d}$ has a term greater than 2, and each positive term of $\underset{\sim}{d}$ is equal to infinitely many other terms then R has uncountably many isolated vertices. (ii) If $\underset{\sim}{d}$ has no term greater than 2, and has infinitely many terms equal to 1, then R has just three isolated vertices. (iii) If $\underset{\sim}{d}$ has no term greater than 2, and has at most one term equal to 1, then the isolated vertices of R form a countably infinite set.*

Proof. The proof is in three parts, corresponding to the cases for $\underset{\sim}{d}$ specified in the Theorem.

(i) Let a > 2 be the value of some term in $\underset{\sim}{d}$, and let $\underset{\sim}{d}'$ be the subsequence comprising all positive terms of $\underset{\sim}{d}$. Corresponding to any nonempty set S of positive integers, we let G be the realization of $\underset{\sim}{d}'$ comprising the disjoint union of the focus $F(\underset{\sim}{d}')$ and infinitely many copies of $H_n(a)$ for each $n \in S$. Note that in G or any of its associates, each infinite component is necessarily the union of a finite component and finitely many infinite components of the form $H_m'(a)$, where each m is in S; morever, the set of values of m arising when all infinite components are taken into account is precisely S. Now let G* be the realization of $\underset{\sim}{d}$ comprising the disjoint union of infinitely many copies of G and each of its associates, together with a set of isolated vertices corresponding to the zero terms of $\underset{\sim}{d}$, if any. Evidently G* is is invariant under switching, so is an isolated vertex of R. Also the induced sub-graphs $H_m'(a)$ in G* permit the identification of the corresponding set S as the set of values of m, so the realizations corresponding to different choices of S are non-isomorphic. Since there are uncountably many choices of S, it follows that R has uncountably many isolated vertices.

(ii) Let M_1 be the focus $F(\underset{\sim}{d})$, let M_2 be $F(\underset{\sim}{d})$ together with infinitely many copies of each of P_∞ and P_∞', and let M_3 be $F(\underset{\sim}{d})$ together with one copy of P_∞'. These three graphs are type $(0,\infty,\infty)$ realizations of $\underset{\sim}{d}$, for in this case it has infinitely many terms equal to 1, infinitely many equal to 2, and no other positive terms. All components of $F(\underline{d})$ are finite paths (possibly including isolated vertices P_1) or cycles (including isolated vertices with one loop C_1). Thus each of M_1, M_2, M_3 can be seen to be invariant under switching, so they are isolated vertices of R. Now suppose G is an isolated vertex of R. If G is not $F(\underline{d})$ itself, it contains $F(\underline{d})$ as an induced subgraph, by Theorem 6. Each finite component of G realizes a finite subsequence of $\underline{d}$, so is in $F(\underline{d})$. Thus G must contain at least one infinite component. Any infinite component of a realization of d must be either P_∞ or P_∞'. If P_∞ is a component of G, it can be switched with a two vertex path P_2 to form two copies of P_∞'. Since G is invariant under switching, it follows that G must already contain infinitely many copies of each of P_∞, P_∞' as components, so G is M_2. Otherwise, G does not have P_∞ as a component, so has at least one component P_∞'. If it has more than one such component, switching between two of them yields a component P_∞ (and a P_2), which is already in G by switching invariance, a contradiction. Thus, G has only one component P_∞', so G is M_3. Thus R has just three isolated vertices.

(iii) First suppose $\underline{d}$ has no term equal to 1. For each $n \in \mathbb{N}$ let N_n be the type $(0, \infty, \infty)$ realization of $\underline{d}$ comprising the focus $F(\underline{d})$ together with n disjoint copies of P_∞. Since the positive terms of $\underline{d}$ are all equal to 2 in this case, all components of $F(\underline{d})$ are finite cycles (including isolated vertices with one loop C_1, and possibly including isolated vertices with no loops C_0). Thus each N_n is switching invariant, and the number of choices of n is countably infinite, so the set of isolated vertices in R is at least countably infinite. In fact, we shall now show that every isolated vertex of R is either N_n for some $n \in \mathbb{N}$, or else is N_∞, the graph comprising $F(\underline{d})$ together with infinitely many disjoint copies of P_∞. Suppose G is an isolated vertex of R which is not N_0, that is, $F(\underline{d})$. Then G contains $F(\underline{d})$ as an induced subgraph, by Theorem 6, and G must contain at least one infinite component. But any infinite component must be a realization of $\underline{2} = (2,2,2,\ldots)$, so must be P_∞. Since the number of components of G can be at most countably infinite, it follows that G is either N_∞ or one of the graphs N_n.

In the case where $\underline{d}$ has exactly one term equal to 1, and all other positive terms equal to 2, a similar argument shows that the isolated vertices of R are precisely the graphs N_n' for each $n \in \mathbb{N}$, together with N_∞', formed by adjoining a single component P_∞' to N_n and N_∞ respectively. □

6. WHEN IS $R(\underline{d};(0,\infty,\infty))$ A NONTRIVIAL PATH?

In this section we are concerned with the case in which $R(\underline{d};(0,\infty,\infty))$ has just one component, so $\underline{d}$ has at most finitely many terms greater than 1 (by Theorem 3). We wish to determine those sequences for which R is a nontrivial path, that is, a path with at least two vertices.

Consider the seven graphs in Figure 9. They cannot occur as subgraphs of a realization of $\underset{\sim}{d}$ if its graph of realizations R is a path. In fact, we can prove a slightly stronger result.

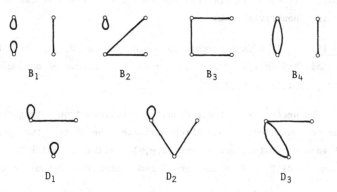

B_1 $\qquad$ B_2 $\qquad$ B_3 $\qquad$ B_4

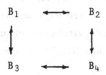

D_1 $\qquad\qquad$ D_2 $\qquad\qquad$ D_3

FIGURE 9 - Seven forbidden subgraphs for realizations in a path

THEOREM 9. *Let $\underset{\sim}{d} \in \mathbf{N}^*$ have at most finitely many terms greater than 1. Any type $(0,\infty,\infty)$ realization of $\underset{\sim}{d}$ which has a subgraph isomorphic to one of B_1, B_2, B_3, B_4, D_1, D_2, D_3, lies on a cycle in $R(\underset{\sim}{d};(0,\infty,\infty))$.*

Proof. If G is a type $(0,\infty,\infty)$ realization of $\underset{\sim}{d}$, it contains only finitely many loops, since loops can only be carried by vertices of degree greater than 1. Thus, if G has a subgraph B isomorphic to one of B_1, B_2, B_3, B_4, a sequence of four switchings beginning on B effects the following transformations:

$$B_1 \longleftrightarrow B_2$$
$$\updownarrow \qquad\qquad \updownarrow$$
$$B_3 \longleftrightarrow B_4$$

The four switchings produce at least three nonisomorphic realizations (distinguished by having different total numbers of loops), and end with G once more. Thus G belongs to a cycle C_3 or C_4 in R in this case.

Similarly, if G contains a subgraph D isomorphic to one of D_1, D_2, D_3, a sequence of three switchings beginning on D will produce a cycle of realizations, because switching effects the following transformations:

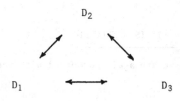

D_2

$D_1 \longleftrightarrow D_3$

Switching produces three nonisomorphic realizations (distinguished by having different total numbers of loops), so G belongs to a cycle C_3 in R in this case. □

This result enables us, in the next theorem, to characterize sequences for which R is a path. It closely parallels Theorem 5, with which it can be combined to specify when R is a nontrivial path.

THEOREM 10.   *If $\underset{\sim}{d}$ is a graphic sequence of type $(0,\infty,\infty)$, then $R(\underset{\sim}{d};(0,\infty,\infty))$ is a path (possibly trivial) precisely when $\underset{\sim}{d}$ has no subsequence equivalent to one which dominates $(2,2,1,1)$ or $(3,2,1)$.*

Proof.   The necessity of these conditions follows from noting that if $\underset{\sim}{d}$ does have a subsequence equivalent to one which dominates one of the two specified sequences then Theorem 1 ensures that it has a type $(0,\infty,\infty)$ realization which has a subgraph isomorphic to one of B_1, D_1. Then Theorems 3 and 9 show that R could not be a path.

For the sufficiency of the stated conditions, we may suppose that $\underset{\sim}{d}$ not only meets these conditions, but has a subsequence equivalent to one which dominates either $(2,2)$ or $(2,1,1)$, for otherwise Theorem 5 shows that R is the trivial path P_1. There are three cases.

(i) $\underset{\sim}{d}$ has at least four positive terms, exactly one of which is greater than 1. Let the largest term of $\underset{\sim}{d}$ be equal to r. The canonical realization of $\underset{\sim}{d}$ has as many loops as possible on the vertex of degree r, with an adjacent vertex of degree 1 just if r is odd, and each other component is either an isolated vertex or a path P_2. Denote this realization by G_0, and let G_i denote the realization in which the vertex of degree r has i fewer loops than in G_0, and correspondingly 2i more adjacent vertices of degree 1. Clearly G_i and G_{i+1} are adjacent vertices of R, so long as $\underset{\sim}{d}$ contains the degree sequence of each, and switching can produce no other adjacencies than these. Since R always contains G_0 and G_1, it is a nontrivial path.

(ii) $\underset{\sim}{d}$ has exactly three positive terms. Parity considerations rule out all but $(2,2,2)$ and $(2n,1,1)$, with $n \geqslant 1$, as the subsequences of positive terms. It is clear in the first case that R is the path P_3, and in the second case the path P_2.

(iii) $\underset{\sim}{d}$ has exactly two positive terms, each of which is greater than 1. Reasoning similar to that in case (1) shows R must be a finite nontrivial path. □

COROLLARY.   *If $R(\underset{\sim}{d};(0,\infty,\infty))$ is a connected graph without cycles, then it is a path (possibly trivial).*

7.   ENDVERTICES WHEN $R(\underset{\sim}{d};(0,\infty,\infty))$ IS CONNECTED

In this final section we are again concerned with the case in which $R(\underset{\sim}{d};(0,\infty,\infty))$

has just one component. Here we wish to determine those sequences for which R has an endvertex, that is, a vertex of degree 1.

Together, Theorems 5 and 10 characterize those sequences for which R is a non-trivial path; in fact, such $\underset{\sim}{d}$ are explicitly specified in the proof of Theorem 10. For each such sequence, R has precisely two endvertices. The following result determines all other cases in which R is a connected graph with at least one endvertex.

THEOREM 11. $R(\underset{\sim}{d};(0,\infty,\infty))$ *is a connected graph with an endvertex, but is not a path, precisely when $\underset{\sim}{d}$ has finitely many positive terms, at least three in number, all of which are even and equal.*

Proof. If $R(\underset{\sim}{d};(0,\infty,\infty))$ is connected, then $\underset{\sim}{d}$ has only finitely many terms greater than 1, by Theorem 3, so any realization of $\underset{\sim}{d}$ has at most a finite number of loops. If R is not a path, $\underset{\sim}{d}$ has a subsequence equivalent to one which dominates either $(2,2,1,1)$ or $(3,2,1)$, by Theorem 10, so $\underset{\sim}{d}$ has at least three positive terms.

Let G be a type $(0,\infty,\infty)$ realization of such a sequence $\underset{\sim}{d}$, and suppose G is an endvertex in R. Then G cannot have two subgraphs (not necessarily disjoint) isomorphic to two of the graphs A_1, A_2, A_3, A_4 (Figure 5). This is because switching on these subgraphs would produce two nonisomorphic realizations adjacent to G in R, since the effect on the total number of loops is different for each of the switchings $A_1 \rightarrow A_2$, $A_2 \rightarrow A_1$, $A_3 \rightarrow A_4$, $A_4 \rightarrow A_3$.

Furthermore, G must have some subgraph A isomorphic to one of the four specified graphs, otherwise G is invariant under switching and R is the trivial graph. We now treat the four cases for A in turn.

(i) $A \cong A_1$. Since $\underset{\sim}{d}$ has at least three positive terms, and no subgraph of G is isomorphic to A_3 or A_4, it follows that G has a subgraph isomorphic to B_4 (Figure 9). The switching $B_4 \rightarrow B_1$ increases the number of loops, while the switching $B_4 \rightarrow B_3$ creates two subgraphs isomorphic to A_3. Thus G has degree at least 2 in R, a contradiction.

(ii) $A \cong A_2$. Since $\underset{\sim}{d}$ has at least three positive terms, and no subgraph of G is isomorphic to A_4, it follows that G has no edges and at least three of its vertices carry loops. Also, every positive term of $\underset{\sim}{d}$ must be even; these terms must all be equal, otherwise the switching $A_2 \rightarrow A_1$ could be used between one vertex of A and either of two vertices of different positive degree, so G would have degree at least 2 in R, a contradiction. It is clear that if all vertices of positive degree in G carry the same number of loops, G cannot have degree greater than 1 in R. This case provides the sequences admitted by the theorem.

(iii) $A \cong A_3$. If G has a subgraph isomorphic to B_3 (Figure 9), the absence of subgraphs isomorphic to A_1 ensures that the switching $B_3 \rightarrow B_4$ creates a realization nonisomorphic to G, while $B_3 \to B_2$ increases the number of loops, so yields yet another nonisomorphic realization. This contradicts the choice of G, so we may suppose no subgraph is isomorphic to B_3. Now it follows that G cannot have two vertices of degree greater than 1: else the absence of subgraphs isomorphic to A_1, A_2, A_4 implies that G would have two disjoint subgraphs isomorphic to A_3; switching between these preserves the number of loops and yields a subgraph isomorphic to B_3, while the switching $A_3 \to A_4$ increases the number of loops, so again G would not be an endvertex in R. It follows that $\underset{\sim}{d}$ has just one term greater than 1; but this contradicts the assumed presence of a subsequence equivalent to one which dominates $(2,2,1,1)$ or $(3,2,1)$.

(iv) $A \cong A_4$. The absence of subgraphs isomorphic to A_1, A_2, A_3 implies that G has just one vertex of degree greater than 1, again contradictory to the assumed dominance properties of $\underset{\sim}{d}$. $\square$

Since the only endvertex realizations admitted by this theorem are those in which all vertices of positive degree carry the same number of loops, each is the unique endvertex in the corresponding graph of realizations. So we conclude with the following result.

COROLLARY. *If* $R(\underset{\sim}{d}; (0,\infty,\infty))$ *is a connected graph with an endvertex, but is not a path, it has precisely one endvertex.*

There are many other fundamental questions about the structure of R which we have not been able to consider in this paper. In particular, if R has more than one component, for which sequences $\underset{\sim}{d}$ does it have a component which is a nontrivial path, and what is the case regarding the presence of endvertices? We plan to take up these questions in a future paper.

REFERENCES

[1]   R.B. Eggleton and D.A. Holton, Graphic sequences, *this volume.*

[2]   J.K. Senior, Partitions and their representative graphs, *Amer. J. Math.*, 73 (1951), 663-689.

Department of Mathematics,
University of Newcastle,
New South Wales, 2308,
Australia.

Department of Mathematics,
University of Melbourne,
Parkville, Vic.   3052,
Australia.

DIVISIBILITY PROPERTIES OF SOME FIBONACCI-TYPE SEQUENCES

A.F. HORADAM, R.P. LOH AND A.G. SHANNON

A generalized Fibonacci-type sequence is defined from a fourth order homogeneous linear recurrence relation, and various divisibility properties are developed. In particular, the notion of a proper divisor is modified to develop formulas for proper divisors in terms of the general terms of the recurrence sequences and various arithmetic functions.

1. THE SEQUENCE $\{A_n(x)\}$

The main sequence of interest is $\{A_n(x)\}$ defined by

(1.1)
$$\begin{cases} A_0(x) = 0, \ A_1(x) = 1, \ A_2(x) = 1, \ A_3(x) = x + 1 \\ \text{and} \\ A_n(x) = x A_{n-2}(x) - A_{n-4}(x) \qquad \text{for } n \geq 4. \end{cases}$$

For unrestricted n, it follows from (1.1) that

(1.2)
$$A_{-n}(x) = -A_n(x) .$$

The auxiliary equation for $\{A_n(x)\}$ is $r^4 - x r^2 + 1 = 0$ which has roots $\pm s, \pm t$ given by

(1.3)
$$\begin{cases} s^2 = \tfrac{1}{2}(x + \sqrt{x^2 - 4}) \\ t^2 = \tfrac{1}{2}(x - \sqrt{x^2 - 4}) \end{cases} \qquad \text{so that } s^2 + t^2 = x, \ s^2 t^2 = 1$$

whence

(1.4)
$$\begin{cases} s = \tfrac{1}{2}(\sqrt{x - 2} + \sqrt{x + 2}) \\ t = \tfrac{1}{2}(\sqrt{x - 2} - \sqrt{x + 2}) \end{cases} \qquad \text{so that } s + t = \sqrt{x - 2}, \ st = -1.$$

From the initial conditions in (1.1) we derive

(1.5)
$$A_{2n}(x) = \frac{s^{2n} - t^{2n}}{s^2 - t^2} .$$

Mathematical induction and the recurrence relation (1.1) lead to

(1.6)
$$A_{2n+2}(x) = A_{2n+1}(x) - A_{2n}(x)$$

whence, by (1.5),

(1.7)
$$A_{2n+1}(x) = \frac{s^{2n}(s^2 + 1) - t^{2n}(t^2 + 1)}{s^2 - t^2}$$

The generating function for $\{A_n(x)\}$ is given by

(1.8)
$$\sum_{n=1}^{\infty} A_n(x) t^n = \frac{t + t^2 + t^3}{1 - xt^2 + t^4}$$

Put $x = 3$ and let α, β be the roots of $r^2 - r - 1 = 0$. Then (1.3) yields

(1.9) $\quad s^2 = \alpha^2, \ t^2 = \beta^2, \ s^2 - t^2 = \alpha - \beta = \sqrt{5}, \ s^2 + 1 = \sqrt{5}\alpha, \ t^2 + 1 = -\sqrt{5}\beta$.

From (1.5), (1.7) and (1.9), we have

(1.10)
$$\begin{cases} A_{2n}(3) = \dfrac{\alpha^{2n} - \beta^{2n}}{\alpha - \beta} = F_{2n} \\[2mm] A_{2n+1}(3) = \alpha^{2n+1} + \beta^{2n+1} = L_{2n+1} \end{cases}$$

in which F_{2n}, L_{2n+1} are the 2n-th Fibonacci number and (2n+1)-th Lucas number respectively. Equation (1.6) is by (1.2) an instance of the well-known result: $F_{k+2} + F_k = L_{k+1}$ which is true for all k. The generating function for $\{A_n(3)\}$ follows immediately from (1.8).

Table 1 shows the first 18 terms of $\{A_n(3)\}$, alternately the Lucas and Fibonacci numbers $L_1, F_2, L_3, F_4, L_5, F_6, L_7, F_8, L_9, \ldots$.

Other examples of $\{A_n(x)\}$ include

$$\left. \begin{matrix} |A_{2n}(-2)| = n \in \mathbb{N} \\[2mm] |A_{2n+1}(-2)| = 1 \end{matrix} \right\} , \text{ obtained directly from (1.1).}$$

Further information about the sequence $\{A_n(x)\}$ may be found in Shannon, Horadam, and Loh [6] where a different notation is used.

2.   PROPER DIVISORS

Vorob'ev [7] in the concluding chapter of his book on Fibonacci numbers refers to the notion of a *proper divisor*. We extend this idea a little as follows:

<u>Definition</u>. For any sequence $\{u_n\}$, $n \geq 1$, where $u_n \in \mathbb{Z}$ or $u_n(x) \in \mathbb{Z}(x)$, the *proper divisor* w_n is the quantity implicitly defined, for $n \geq 1$, by $w_1 = u_1$ and $w_n = \max\{d: d|u_n$ and g.c.d.$(d,w_m) = 1$ for every $m < n\}$.

(Strictly speaking, the second equation is all that is necessary here, since for $n = 1$ its g.c.d. condition is vacuous and so $w_1 = u_1$ follows.)

Proper divisors w_n for the sequence of integers $\{A_n(3)\}$ are the integers listed in Table 1. (Recall (1.10).)

n	1	2	3	4	5	6	7	8	9	10	11	12	13	14	15	16	17	18
$A_n(3)$	1	1	4	3	11	8	29	21	76	55	199	144	521	377	1364	987	3571	2584
w_n	1	1	4	3	11	–	29	7	19	5	199	–	521	13	31	47	3571	17

Table 1. Proper divisors for $\{A_n(3)\}$

Proper divisors for the sequence of polynomials $\{A_n(x)\}$ are shown in Table 2. These proper divisors $w_n(x)$ are monic polynomials (over the integers).

n	$A_n(x)$	$w_n(x)$
3	$x+1$	$x+1$
4	x	x
5	x^2+x-1	x^2+x-1
6	x^2-1	$x-1$
7	x^3+x^2-2x-1	x^3+x^2-2x-1
8	x^3-2x	x^2-2
9	$x^4+x^3-3x^2-2x+1$	x^3-3x+1
10	x^4-3x^2+1	x^2-x-1
11	$x^5+x^4-4x^3-3x^2+3x+1$	$x^5+x^4-4x^3-3x^2+3x+1$
12	x^5-4x^3+3x	x^2-3

Table 2. Proper divisors for $\{A_n(x)\}$

From the definition of proper divisors we have that for $\{A_n(x)\}$:

$$A_p(x) = w_p(x)\, w_1(x)$$

$$A_{2p}(x) = w_{2p}(x)\, w_p(x)\, w_2(x)\, w_1(x)$$

$$A_{3p}(x) = w_{3p}(x)\, w_p(x)\, w_3(x)\, w_1(x)$$

$$\cdots\cdots\cdots\cdots\cdots\cdots\cdots\cdots\cdots\cdots\cdots\cdots\cdots$$

in which $w_2(x) = w_1(x) = 1$ for notational convenience.

Hence,

(2.1)
$$A_n(x) = \prod_{d|n} w_d(x)$$

<u>Theorem 1.</u> $w_n(x) = \prod_{d|n} (A_d(x))^{\mu(n/d)}$ *where μ is the Möbius function.*

<u>Proof</u>. Taking logarithms in (2.1) we obtain

$$\ln A_n(x) = \sum_{d|n} \ln w_d(x)$$

which becomes, with the Möbius inversion formula,

$$\ln w_n(x) = \sum_{d|n} \mu(n/d)\, \ln A_d(x)$$

i.e.
$$w_n(x) = \prod_{d|n} (A_d(x))^{\mu(n/d)} \quad \text{as required.}$$

As an example,

$$w_{12}(x) = (A_3(x))^{\mu(4)} (A_4(x))^{\mu(3)} (A_6(x))^{\mu(2)} (A_{12}(x))^{\mu(1)}$$

$$= (x + 1)^0 \, x^{-1} (x^2 - 1)^{-1} (x^5 - 4x^3 + 3x)^1$$

$$= x^2 - 3.$$

Another approach to Theorem 1 is through cyclotomic polynomials.

3. PROPERTIES OF PROPER DIVISORS

Let $n = \prod_{i=1}^{m} p_i^{\alpha_i}$ where p_i are distinct primes, and let $\nu(n) = \sum_{i=1}^{m} \alpha_i$ be the number of prime factors of n, counted with multiplicity. Further, let

$$\varepsilon(n) = (-1)^{\nu(n)}.$$

Then

Theorem 2. $w_n(x) = \left(\prod_{d \in S_0} A_d(x) \Big/ \prod_{d \in S_1} A_d(x) \right)^{\varepsilon(n)}$, *where the sets* S_i *comprise all positive divisors* d *of* n *such that* n/d *is squarefree, and* $\nu(d) \equiv i \bmod(2)$ *for* $i = 0, 1$.

Proof. Write $d = \prod_{i=1}^{m} p_i^{\beta_i}$ $(\beta_i \le \alpha_i)$.

Then $n/d = \prod_{i=1}^{m} p_i^{\alpha_i - \beta_i}$

and $\mu(n/d) = (-1)^{\Sigma(\alpha_i - \beta_i)}$ $(0 \le \alpha_i - \beta_i \le 1)$,

since $\mu(n/d) = 0$ if $\alpha_i - \beta_i \ge 2$.

From Theorem 1 we have, using the specified notation,

$$w_n(x) = \prod_{d|n} (A_d(x))^{(-1)^{\Sigma(\alpha_i - \beta_i)}}$$

$$= \prod_{d|n} (A_d(x))^{\varepsilon(n) \cdot (-1)^{\Sigma \beta_i}}$$

$$= \prod_{d|n} (A_d(x))^{\varepsilon(n)\varepsilon(d)}$$

$$= \left(\prod_{d \in S_0} A_d(x) \Big/ \prod_{d \in S_1} A_d(x) \right)^{\varepsilon(n)}$$

since $\varepsilon(d)$ will be positive or negative according as $d \in S_0$ or $d \in S_1$, respectively.

For example,
$$w_{60}(x) = \frac{A_4(x) \; A_6(x) \; A_{10}(x) \; A_{60}(x)}{A_2(x) \; A_{12}(x) \; A_{20}(x) \; A_{30}(x)}$$

since $\qquad \nu(60) = 4 \qquad$ (i.e. $\epsilon(60) = 1$)

with $\qquad S_0 = \{4, 6, 10, 60\}$,

and $\qquad S_1 = \{2, 12, 20, 30\}$.

Note that $\mu(60/d) = 0$ for $d = 3, 5, 15$.

Corollary 1. $\qquad w_{p^n}(x) = A_{p^n}(x) \big/ A_{p^{n-1}}(x)$.

Corollary 2. $\qquad w_{2^k}^2(x) - w_{2^{k+1}}(x) = 2 \qquad (k \geq 2)$.

The proof of Corollary 2 uses Corollary 1 and (1.5).

Corollary 2 may be illustrated by choosing $k = 2$, whence, by Table 2,
$$w_4^2(x) - w_8(x) = x^2 - (x^2 - 2) = 2.$$

Other results, such as

Corollary 3. $\qquad w_{2^{k+1}}^2(x) - w_{2^k \cdot 6}(x) = 3$

can be similarly proved. Putting $k = 1$ in Corollary 3 and using Table 2, we see that
$$w_4^2(x) - w_{12}(x) = x^2 - (x^2 - 3) = 3.$$

4. GENERALIZED PELL NUMBERS

Consider the *forward shift operator* E:

(4.1) $\qquad E \, A_n(x) = A_{n+1}(x)$.

Then (1.1), along with (1.3), can be written as
$$(E^2 - s^2)(E^2 - t^2) \, A_n(x) = 0$$

i.e. $\qquad (E - s)(E - t)(E + s)(E + t) \, A_n(x) = 0$

so

(4.2) $\qquad (E - s)(E - t) \, \phi_n(x) = 0$

where
$$\phi_n(x) = (E^2 + (s + t)E + st) \, A_n(x)$$

i.e.

(4.3) $\qquad \phi_n(x) = A_{n+2}(x) + M \, A_{n+1}(x) - A_n(x)$

where, by (1.4), $\qquad M = s + t = \sqrt{x - 2}$.

Equation (4.2) can, with (1.4), be rewritten as

(4.4) $\qquad \phi_{n+2}(x) = M \, \phi_{n+1}(x) + \phi_n(x)$

which is the usual form of the Pellian recurrence relation. Consequently, we may call $\{\phi_n(x)\}$ a *generalized Pell sequence* and $\phi_n(x)$ *generalized Pell numbers.*

Equation (4.3) relates *generalized Pell numbers* to numbers of the sequence $\{A_n(x)\}$.

When $M = 2$ in (4.4) (i.e., $x = 6$, $s = 1 + \sqrt{2}$, $t = 1 - \sqrt{2}$ in (1.4)), we have the most common form which is used to generate the ordinary Pell sequence of numbers $\{P_n\} = \{1, 2, 5, 12, 29, 70, 169, 408, \ldots\}$, $n \geq 1$, defined (Horadam [5]) by

$$(4.5) \qquad P_{n+2} = 2P_{n+1} + P_n \quad \text{with} \quad P_0 = 0, \quad P_1 = 1.$$

Furthermore, the explicit form of P_n is

$$(4.6) \qquad P_n = \frac{s^n - t^n}{s - t} \qquad (s - t = 2\sqrt{2} \text{ from } (1.4)).$$

For unrestricted n, (4.6) yields

$$(4.7) \qquad P_{-n} = (-1)^{n+1} P_n .$$

Elements of the sequences $\{A_n(x)\}$, $\{\phi_n(x)\}$ and $\{P_n\}$ are related thus, as may be demonstrated:

$$(4.8) \qquad \begin{cases} \phi_{2n+1}(x) = P_{2n+1} + A_{2n+3}(x) \\ \phi_{2n}(x) = P_{2n} - 2P_{2n+2} + A_{2n+3}(x) . \end{cases}$$

Table 3 shows the first few numbers in the sequences $\{A_n(6)\}$ and $\{\phi_n(6)\}$ which are obtained from recurrence relations (1.1) and (4.3), and are confirmed by the recurrence relation (4.4).

n	1	2	3	4	5	6	7
$A_n(6)$	1	1	7	6	41	35	239
$\phi_n(6)$	8	19	46	111	268	647	1562

Table 3. $A_n(6)$ and $\phi_n(6)$

Observe that $\phi_0(6) = P_0 - 2P_2 + A_3(6) = 3$ by (4.5), (4.8) and Table 3.

Values of $\phi_{-n}(6)$ may by calculated from (4.8) in conjunction with (1.1), (1.2), (4.7) and (4.8).

Theorem 3. *If, in (4.4), $M = 2N$ (even), $n \geq 0$, $\phi_0(N) = 1$, $\phi_1(N) = N$, then $(\phi_{2n}^2(N) - 1)/(N^2 + 1)$ is a perfect square.*

Proof. The explicit form of $\phi_n(N)$ is, by the usual method,

$$\phi_n(N) = \tfrac{1}{2}(c^n + d^n)$$

where $c = N + \sqrt{N^2 + 1}$ and $d = N - \sqrt{N^2 + 1}$, so $c - d = 2\sqrt{N^2 + 1}$ and $cd = -1$.

Hence
$$\frac{\phi_{2n}^2(N) - 1}{N^2 + 1} = \frac{(c^{2n} + d^{2n})^2 - 4}{4(N^2 + 1)}$$

$$= \left(\frac{c^{2n} - d^{2n}}{c - d}\right)^2$$

and $(c^{2n} - d^{2n})/(c - d)$ is an integer for non-negative integer n.

For example $\dfrac{\phi_4^2(N) - 1}{N^2 + 1} = [4N(2N^2 + 1)]^2$ when $n = 2$.

When $N = 1$ in Theorem 3, we have the sequence $\{\phi_n\}$ say:

ϕ_0	ϕ_1	ϕ_2	ϕ_3	ϕ_4	ϕ_5	ϕ_6	ϕ_7	ϕ_8	ϕ_9 ...
1	1	3	7	17	41	99	239	577	1393 ...

whence

$$\begin{cases} \phi_{2n+1} = A_{2n+1} \\ \phi_{2n} = P_{2n} + P_{2n-1} \end{cases} (6)$$

and

$$\frac{\phi_{2n}^2 - 1}{2} = P_{2n}^2 \quad \text{(by (4.6) since } c = s, \ d = t \text{ when } N = 1).$$

In the illustrative example above, when $n = 2$ we have $\dfrac{\phi_4^2 - 1}{2} = \dfrac{289 - 1}{2} = 144 = 12^2 = P_4^2$.

5. A FIBONACCI-TYPE SEQUENCE

Consider the sequence $\{Q_n(N)\}$ where $N \neq 2$ is an integer:

(5.1)
$$\begin{cases} Q_{n+2}(N) = N\,Q_{n+1}(N) - Q_n(N) & (n \geq 1) \\ Q_1(N) = Q_2(N) = 1. \end{cases}$$

The first few numbers of this sequence are given in Table 4:

n	$Q_n(N)$
1	1
2	1
3	$N - 1$
4	$N^2 - N - 1$
5	$N^3 - N^2 - 2N + 1$
6	$N^4 - N^3 - 3N^2 + 2N + 1$
7	$N^5 - N^4 - 4N^3 + 3N^2 + 3N - 1$
8	$N^6 - N^5 - 5N^4 + 4N^3 + 6N^2 - 3N - 1$
9	$N^7 - N^6 - 6N^5 + 5N^4 + 10N^3 - 6N^2 - 4N + 1$

Table 4.

An interesting factorization result arises from these number, namely:

Theorem 4. $\qquad Q_{n+1}(N) - 1 = (N - 2) A_n(N) A_{n-1}(N) \qquad (n \geq 2, N \neq 2)$.

Proof. $\qquad$ We use induction. The result is obvious when $n = 2,3$. Assume the result is true for $n = 4, 5, \ldots, k+1$. Then

$$
\begin{aligned}
Q_{k+2}(N) &= N\,Q_{k+1}(N) - Q_k(N) && \text{by (5.1)} \\
&= N\,Q_{k+1}(N) - N\,Q_{k-1}(N) + Q_{k-2}(N) && \text{by (5.1)} \\
&= N(N-2)\,A_k(N)\,A_{k-1}(N) + N - N(N-2)\,A_{k-2}(N)\,A_{k-3}(N) \\
&\quad - N + (N-2)\,A_{k-3}(N)\,A_{k-4}(N) + 1 && \text{by hypothesis} \\
&= N(N-2)\,A_k(N)\,A_{k-1}(N) + (N-2)\,A_{k-3}(N)\Big[A_{k-4}(N) - N\,A_{k-2}(N)\Big] + 1 \\
&= N(N-2)\,A_k(N)\,A_{k-1}(N) - (N-2)\,A_{k-3}(N)\,A_k(N) + 1 && \text{by (1.1)} \\
&= (N-2)\,A_k(N)\Big[N\,A_{k-1}(N) - A_{k-3}(N)\Big] + 1 \\
&= (N-2)\,A_k(N)\,A_{k+1}(N) + 1 && \text{by (1.1)}
\end{aligned}
$$

as required.

For example, $n = 7$ in Theorem 4 gives, with the help of Table 2,

$$Q_8(N) - 1 = (N-2)\,A_7(N)\,A_6(N) = (N-2)(N^2 - 1)(N^3 + N^2 - 2N - 1).$$

Further, $N = 3$ in this example yields, with (1.10),

$$F_{13} - 1 = L_7 F_6 \qquad (= 232 = 29 \times 8).$$

since the numbers $Q_n(3)$ are certain Fibonacci numbers.

Induction also leads to the result

(5.2) $\qquad Q_n(N) = \displaystyle\sum_{j=0}^{[(n-2)/2]} (-1)^j \binom{n-j-2}{j} N^{n-2j-2} - \sum_{j=0}^{[(n-3)/2]} \binom{n-j-3}{j} N^{n-2j-3}$

which is a generalization of (2.8) of Barakat [1].

When $N = 3$, (5.2) reduces to

$$Q_n(3) = F_{2n-2} - F_{2n-4}.$$

For example, $Q_5(3) = 13 = F_8 - F_6 \quad (= 21 - 8)$.

A basic relationship amongst the elements of $\{Q_n(N)\}$ is

(5.3) $\qquad Q_{n+1}(N)\,Q_{n-1}(N) - Q_n^2(N) = N - 2$

which is a particular case of the general result for the sequence $\{w_n(a,b;p,q)\}$ in Horadam [4] where $a = 1, b = 1, p = N, q = 1$. (Equation (5.3) is the analogue for $\{Q_n(N)\}$ of the well-known Simson result for Fibonacci numbers: $F_{n+1} F_{n-1} - F_n^2 = (-1)^n$,

$1 \geq 1$.)

Following the approach of Hoggatt and Bicknell [3] for Fibonacci polynomials and letting e^z and e^{-z} be the roots of the auxiliary equation $r^2 - Nr + 1 = 0$ associated with the recurrence relation (5.1), we obtain

(5.4)
$$Q_{n+1}(N) = \frac{\cosh \frac{1}{2}(2n-1)z}{\cosh \frac{1}{2}z}$$

where $2 \cosh z = N$, $2 \sinh z = \sqrt{N^2 - 4}$.

Clearly, further results may be developed involving the sequences under consideration, e.g.

(5.5)
$$Q_n(6) = P_{2n-3} \qquad (n \geq 2),$$
and

(5.6)
$$Q_n'(6) = P_{2n-2} \qquad (n \geq 1)$$

if we define $Q_n'(N)$ as for $Q_n(N)$ in (5.1) but with the initial conditions $Q_1'(N) = 0$, $Q_2'(N) = 2$.

The theory for $Q_n(N)$ and $Q_n'(N)$ extends to negative values of n.

CONCLUSION

It is of interest to note ways in which this work can be further extended. When $N = 2$, the Pell equation can be readily related to the Diophantine equations

$$x^2 - 2y^2 = \pm 1$$

because of the simple continued fraction expansion of $\sqrt{2}$, namely,

$$\sqrt{2} = [1, \dot{2}] .$$

Bernstein [2] has shown how it can be further developed by considering the surd

$$\sqrt{m} = [b_0, \dot{b}_1, b_2, \ldots, b_{n-1}, 2\dot{b}_0]$$

and the recurrence relation

$$P_{j+2} = b_j P_{j+1} + P_j \qquad (j > 0),$$

where b_j is the jth partial quotient of the continued fraction, and with suitable initial conditions. Bernstein has generalized this further by using the Jacobi-Perron algorithm to accommodate linear recurrence relations of order higher than two.

ACKNOWLEDGEMENTS

The authors wish to express their indebtedness to Dr. R.B. Eggleton, University of Newcastle, whose valuable comments have improved the presentation of this article.

The first and last named authors did some of this work whilst they were on study leave in England at the School of Mathematics and Physics, University of East Anglia, Norwich, and Linacre College, Oxford, respectively. Facilities made available to them at these institutions are gratefully acknowledged.

REFERENCES

[1]   R. Barakat,   'The matrix operator e^x and the Lucas polynomials', *J. Math. and Physics*, 43 (1964), 332 - 335.

[2]   L. Bernstein, 'Explicit solutions of pyramidal Diophantine equations', *Canad. Math. Bull.*, 15 (1972), 177 - 184.

[3]   V.E. Hoggatt, Jr. and Marjorie Bicknell, 'Roots of Fibonacci polynomials', *Fibonacci Quart.*, 11 (1973), 271 - 274.

[4]   A.F. Horadam, 'Basic properties of a certain generalized sequence of numbers', *Fibonacci Quart.*, 11 (1965), 161 - 176.

[5]   A.F. Horadam, 'Pell identities', *Fibonacci Quart.*, 9 (1971), 245 - 252, 263.

[6]   A.G. Shannon, A.F. Horadam and R.P. Loh, 'Proper divisors of Lenakel numbers', *Math. Stud.*, (to appear).

[7]   N.N. Vorob'ev, *Fibonacci Numbers*. (Pergamon, Oxford, 1961.)

Department of Mathematics,
University of New England,
ARMIDALE N.S.W. 2351

Department of Applied Mathematics,
University of Sydney,
SYDNEY N.S.W. 2006

School of Mathematical Sciences,
N.S.W. Institute of Technology,
BROADWAY N.S.W. 2007

INTERLACED TREES: A CLASS OF GRACEFUL TREES

K.M. KOH, T. TAN

AND D.G. ROGERS

Ringel conjectured that every tree has a graceful valuation. While this conjecture remains unsettled, it is apparent, from examples, that some trees have graceful valuations with additional properties which allow larger trees with graceful valuations to be constructed from them. We investigate here one such class of graceful trees.

INTRODUCTION

Let T denote, throughout this section, a tree on n vertices. A *valuation* θ of T is a bijection between the vertex set of T and the set $\{1, \ldots, n\}$. The vertex of T for which $\theta(b) = 1$ is called the *base* of the valuation and $A(b)$ is the set of vertices of T adjacent to b. For vertices u and v in T, we write $(u,v) = d(v,u)$ for the length of (number of edges in) the shortest path in T between and v (with $d(u,u) = 0$) and, in particular, we set $d(v) = d(b,v)$. (If we wish o emphasize the tree T we are considering, we write $d_T(u,v)$ or $d_T(v)$ and similarly ith our other notation.) For a vertex v in T, let $\mathcal{S}(v)$ be the set of vertices of T (including v) for which $d(u,v)$ is even and let $s(v)$ be the number of ertices in $\mathcal{S}(v)$. If T is a tree having a graceful valuation θ with base b then he *size* s of T under θ is $s = s(b)$. A valuation θ of T with base b is a *arity valuation* if it induces, by restriction, a bijection between $\mathcal{S}(b)$ and the set 1, $\ldots$, s$\}$. An edge of T with endpoints u and v is denoted by $<u,v>$. If λ s any numerical labelling of the vertices of T, an edge $<u,v>$ of T carries under the *weight* $\omega(u,v) = \omega_\lambda(u,v) = |\lambda(u) - \lambda(v)|$. The valuation θ is *graceful* if distinct edges carry distinct weights, that is ω is a bijection between the edge set of and the set $\{1, \ldots, n - 1\}$. T is a *graceful tree* if it has a graceful valuation. ote that if θ is a graceful valuation then so is θ^+ where

$$\theta^+(v) = n + 1 - \theta(v). \tag{1}$$

Ringel conjectured, in 1963, that every tree is graceful, the term "graceful" being ue, however, to Golomb [4]. While several classes of trees have been shown to be raceful, Ringel's conjecture remains unsettled. Cahit gave wider circulation to this conjecture by raising, as a research problem [1], the more specific question: *are all complete binary trees graceful?* Although an affirmative answer to Cahit's question was already contained in a paper of Stanton and Zarnke [12], his note provided further stimulus to work on the subject, resulting recently in other alternative or duplicate solutions together with generalizations and refinements (including work of the present authors [7,8,9,10]: other references are collected in [5]). Related research and discussions are contained in [2,3,6].

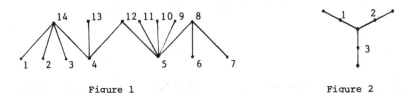

Figure 1   Figure 2

Several of the graceful valuations illustrated in Cahit's note have an additional property which we now single out for consideration. An *interlaced valuation* θ of T is a graceful valuation which is also a parity valuation. T is an *interlaced tree* if it has an interlaced valuation. Such valuations have a number of elementary but useful features. Thus if T has size s under θ and $\theta(u) = s$, then u must be adjacent to a vertex v of T such that $|\theta(u) - \theta(v)| = 1$ since otherwise no edge of T would carry weight one. Further, if a labelling λ of the vertices of T is defined by

$$\lambda(v) = \theta(v), \quad \theta(v) \le s \; ; \; = \theta(v) + k, \quad s < \theta(v),$$

then the edges of T carry under λ all the weights in the range k to $k + n - 1$. Also if θ is an interlaced valuation of T then so is θ^+ given by (1) as is θ' where

$$\theta'(v) = s + 1 - \theta(v), \quad \theta(v) \le s \; ; \; = n + s + 1 - \theta(v), \quad s < \theta(v). \qquad (2)$$

An interlaced valuation is a special type of what Rosa [11] has called an α-*valuation:* a graceful valuation θ of T is an α-*valuation* if and only if there is a number r such that whenever u and v are adjacent vertices in T either $\theta(u) \le \theta(v)$ or $\theta(v) \le r < \theta(u)$ — thus if θ is interlaced, $r = s$ satisfies this requirement.

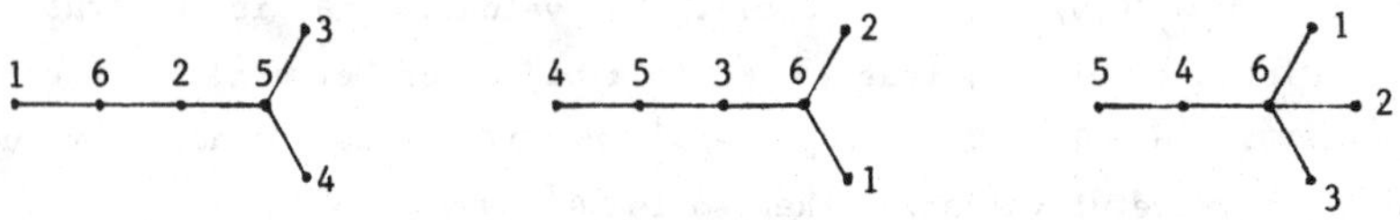

Figure 3

Now the graceful valuations given by Cahit suggest the conjecture that all complete binary trees are interlaced and this was established by an algorithm in [10] which we discuss in more detail in §4. Secondly, trees (called *caterpillars*) in which the vertices either lie on a central strand (chain) or are adjacent to vertices on the central strand are interlaced — a typical example is illustrated in Figure (1). Not all trees, however, are interlaced. For example, the tree in Figure (2) is not inter-laced since if it were we may suppose, by symmetry and the remark concerning θ' in the previous paragraph, that the interlaced valuation begins as shown and it is then easy to see that it cannot be completed. It is useful for subsequent developments to note here that there may be distinct interlaced trees on the same number of vertices with the same size (see Figure (3)) and that the same tree may have two interlaced valuations θ and ϕ with θ, θ^+, θ', ϕ, ϕ^+ and ϕ' all distinct (see Figure (4)).

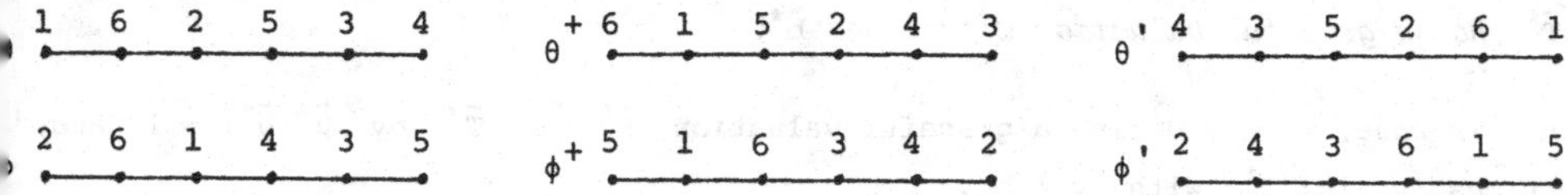

θ 1   6   2   5   3   4 θ^+ 6   1   5   2   4   3 θ' 4   3   5   2   6   1

ϕ 2   6   1   4   3   5 ϕ^+ 5   1   6   3   4   2 ϕ' 2   4   3   6   1   5

Figure 4

Interlaced trees may be put together in various ways to form further graceful trees which in some circumstances are also interlaced. In §3, we give some general results in this direction and show by a number of examples the sort of trees which may be obtained in this way. On the other hand, the classification of interlaced trees is left open and, in view of our positive results (§3,4) and Figure (2), may be difficult. We begin, in §2, by showing how interlaced trees may be used in the constructions of [7,8,9] to obtain larger graceful trees (which, however, are not themselves interlaced).

2.   SOME GRACEFUL CONSTRUCTIONS

Let T be a tree on n vertices with graceful valuation θ having base b. Let T_i, $1 \leq i \leq p$, be p disjoint isomorphic copies of T, let v_i be the image in T_i of the vertex v in T and let $\bar{T}$ be the tree obtained from the T_i by making the identifications $b_1 = b_2 = \ldots = b_p = \bar{b}$. Then the following theorem gives a sufficient condition for $\bar{T}$ to be graceful.

THEOREM 1 [8]. *If* $\{n - \theta(v) : v \in A(b)\} \subseteq \{0\} \cup \{\theta(v) - 1 : v \in A(b)\}$, *then* $\bar{T}$ *has a graceful valuation with base* $\bar{b}$.

Indeed one graceful valuation $\bar{\theta}$ of $\bar{T}$ is given by $\bar{\theta}(\bar{b}) = 1$ and, for $1 \leq i \leq p$ with $v_i \neq b_i$,

$$\bar{\theta}(v_i) = \begin{cases} \theta(v) + (i - 1)(n - 1), & v \in \mathcal{S}_T(b) \\ \theta(v) + (p - i)(n - 1), & v \in \mathcal{S}_T(b) \end{cases} . \tag{3}$$

This theorem is an extension of one given earlier in [7] where the condition on $A(b)$ holds automatically as b has valence one (see [8]). As a corollary, we may deduce that every complete m-any tree is graceful [7,8].

The form of the definition (3) suggests replacing isomorphic copies of one tree by (possibly) distinct disjoint interlaced trees. Thus let T_i^* , $1 \leq i \leq p$, be p disjoint trees, each on n vertices, having respectively interlaced valuations θ_i^* with bases b_i^* but common size $s_i^* = s^*$. Similarly let $\bar{T}^*$ be the tree obtained from the T_i^* by making the identifications $b_1^* = b_2^* \ldots = b_p^* = \bar{b}^*$. Then, corresponding to Theorem 1, we have:-

THEOREM 2. If $\{\theta_i^*(v) : v \in A(b_i^*)\}$ *is independent of* i *and for each* i, $1 \le i \le p$,

$$\{n - \theta_i^*(v) : v \in A(b_i^*)\} \subseteq \{0\} \cup \{\theta_i^*(v) - 1 : v \in A(b_i^*)\} ,$$

then $\bar{T}^*$ *has a graceful valuation with base* $\bar{b}^*$.

In this case, we may define a graceful valuation $\bar{\theta}^*$ of $\bar{T}^*$ by $\bar{\theta}^*(\bar{b}^*) = 1$ and, for vertices v in T_i^* with $v \ne b_i$,

$$\bar{\theta}^*(v) = \begin{cases} \theta_i^*(v) + (i-1)(n-1), & \theta_i^*(v) \le s^* , \\ \theta_i^*(v) + (p-i)(n-1), & s^* < \theta_i^*(v). \end{cases}$$

That $\bar{\theta}^*$ is a graceful valuation of $\bar{T}^*$ follows in exactly the same way as for the proof of Theorem 1. (However, since each θ_i^* is interlaced, the weights under $\bar{\theta}^*$ of the edges of T_i^*, other than those incident at $b^* = b_i$, come in consecutive runs, the gaps between these runs being filled by the weights on the edges incident at b^*. This slightly simplifies the proof of Theorem 2 and similar simplifications are also possible in the proofs of Theorems 4 and 6.) An illustration of the construction used in Theorem 2, based on the trees of Figure (3), is shown in Figure (5).

Interlaced trees may be used in a second generalization [9] of the results in [7]. Thus let T, T_i and T_i^*, $1 \le i \le p$, be as before and let c and c_i^* be the vertices in T and T_i^* respectively for which $\theta(c) = n = \theta_i^*(c_i^*)$. In addition, let J be a tree on $(p+1)$ vertices w_r, $0 \le r \le p$, with a graceful valuation ψ having base w_0 and let $J \circ T(\text{resp. } J \circ T^*)$ be the tree obtained from J by adjoining, for each i, $1 \le i \le p$, the tree $T_i(\text{resp. } T_i^*)$ at the vertex w_i with the identification $w_i = c_i(\text{resp. } w_i = c_i^*)$. Then we have the pair of companion results:-

THEOREM 3 [9]. $J \circ T$ *has a graceful valuation with base* w_0.

THEOREM 4. $J \circ T^*$ *has a graceful valuation with base* w_0.

Again the proof, for both theorems, is a matter of defining the appropriate valuations. Graceful valuations $\hat{\theta}$ and $\hat{\theta}^*$ of $J \circ T$ and $J \circ T^*$ respectively are given by, for $1 \le i \le p$,

$$\hat{\theta}(v_i) = \begin{cases} 1 + \theta(v) + (\psi(w_i) - 2)n, & v \notin \mathcal{S}_\pi(b) , \\ 1 + \theta(v) + (p + 1 - \psi(w_i))n, & v \in \mathcal{S}_\pi(b) ; \end{cases}$$

and, with v in T_i^*,

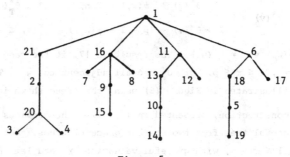

Figure 5

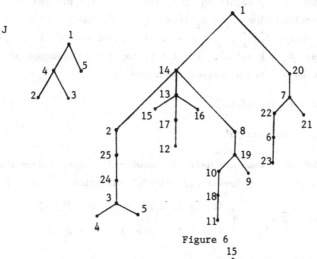

Figure 6

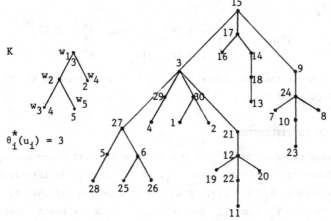

Figure 7

$$\hat{\theta}^*(v) = \begin{cases} 1 + \theta_i^*(v) + (\psi(w_i) - 2)n, & s^* < \theta_T^*(v) , \\ 1 + \theta_i^*(v) + (p + 1 - \psi(w_i))n, & \theta_i^*(v) \le s^* , \end{cases}$$

where also $\hat{\theta}(w_0) = 1 = \hat{\theta}^*(w_\theta)$. The result in [7] is the case of Theorem 3 in which the vertices w_r, $1 \le r \le p$, of J are all adjacent to w_0. The construction for Theorem 4 is illustrated in Figure (6) using the trees shown in Figure (3).

A third construction, presented in [8], like the previous one, enables us to assemble a graceful tree from two smaller graceful trees. Let K be a tree on p vertices x_r, $1 \le r \le p$, with graceful valuation χ and let $K \Delta T$(resp. $K \Delta T^*$) be the tree obtained from K by adjoining, for each i, $1 \le i \le p$, the tree T_i(resp. T_i^*) at x_i with the identification $x_i = u_i$ (resp. $x_i = u_i^*$) where u is an arbitrary vertex of T (resp. where $\theta_i^*(u_i^*)$, $1 \le i \le p$, is independent of i but is otherwise arbitrary – note that $\theta_i(u_i) = \theta(u)$, $1 \le i \le p$, is also independent of i but arbitrary). Then we have again a pair of results:-

THEOREM 5 [8]. $K \Delta T$ is graceful.

THEOREM 6. $K \Delta T^*$ is graceful.

In this case the valuations are defined somewhat differently from the two previous cases. For $K \Delta T^*$, we have respectively graceful valuations $\widetilde{\theta}$ and $\widetilde{\theta}^*$ given by:-

$$\widetilde{\theta}(v_i) = \begin{cases} \theta(v) + (\chi(w_i) - 1)n, & v \in \mathscr{P}_T(b) , \\ \theta(v) + (p - \chi \ell w_i))n, & v \notin \mathscr{P}_T(b) ; \end{cases}$$

and, with v in T_i^*,

$$\widetilde{\theta}^*(v) = \begin{cases} \theta_i^*(v) + (\chi(w_i) - 1)n, & \theta_i^*(v) \le s^*, \\ \theta_i^*(v) + (p - \chi(w_i))n, & s^* < \theta_i^*(v) . \end{cases}$$

An example of the construction for Theorem 6, based as before on the trees of Figure (3), is shown in Figure (7).

3.   SOME INTERLACED CONSTRUCTIONS

In the following two results, T_i, i = 1,2, are disjoint trees on n_i vertices having graceful valuations θ_i with bases b_i. Also, if θ_i is interlaced, then s_i is the size of T_i under θ_i, i = 1,2. The results show that if the θ_i are interlaced, we may amalgamate T_1 and T_2 to form trees which are again interlaced. We illustrate this in a number of cases where the resulting trees have a regular pattern (see Figures 8-13).

THEOREM 7.   Suppose that θ_1 is interlaced and the vertex x in T_1 is such that $\theta(x) = s_1$. Let T be the tree obtained from T_1 and T_2 by identifying x in T_1 with b_2 in T_2. Then T is graceful. Moreover, if θ_2 is interlaced, then T is an interlaced tree.

PROOF: Define θ on the vertex set of T by:-

$$\theta(v) = \begin{cases} \theta_1(v) & v \in \mathscr{S}_{T_1}(b_1) , \\ \theta_1(v) + n_2 - 1, & v \notin \mathscr{S}_{T_1}(b_1), \\ \theta_2(v) + -1 & , \quad v \text{ in } T_2 . \end{cases}$$

Then θ is a valuation of T and under θ the edges of T_1 take all the weights in the range n_2 to $n_1 + n_2 - 2$ while those of T_2 take all the weights in the range 1 to $n_2 - 1$. If, in addition, θ_2 is interlaced, then θ is also interlaced, as the bases b_1 and $b_2 = x$ are an even distance apart.

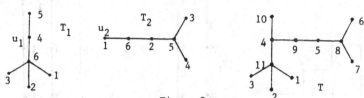

Figure 8

THEOREM 8.   *Suppose that both θ_1 and θ_2 are interlaced. Suppose further that there are vertices u_i in T_i, $i = 1,2$, such that either*

(i) $\theta_1(u_1) - \theta_2(u_2) = s_1 < \theta_1(u_1)$;

or

(ii) $n_2 + \theta_1(u_1) - \theta_2(u_2) = s_1 \geq \theta_1(u_1)$.

Let T be the tree obtained from T_1 and T_2 by joining them at u_1 and u_2 by a new edge. Then T is graceful. Further if, in the two cases (i) $\theta_2(u_2) \leq s_2$; or (ii) $\theta_2(u_2) > s_2$, then T is interlaced.

PROOF   Define θ on the vertex set of T by:-

$$\theta(v) = \begin{cases} \theta_1(v) & , \quad v \in \mathscr{S}_{T_1}(b) , \\ \theta_1(v) + n_2, & v \notin \mathscr{S}_{T_1}(b) , \\ \theta_2(v) + s_1, & v \text{ in } T_2. \end{cases}$$

So θ is a valuation of T. The edges of T_2 carry under θ all the weights 1 to $n_2 - 1$ while those of T_1 carry all the weights $n_2 + 1$ to $n_1 + n_2 - 1$. Moreover, in case (i) ,

$$|\theta(u_1) - \theta(u_2)| = \theta(u_1) - \theta(u_2) = \theta_1(u_1) + n_2 - (\theta_2(u_2) + s_1) = n_2 ,$$

while, in case (ii),

$$|\theta(u_1) - \theta(u_2)| = \theta(u_2) - \theta(u_1) = (\theta_2(u_2) + s_1) - \theta_1(u_1) = n_2 ,$$

so that, in either case, the new edge carries the weight n_2. Hence the edges of T carry all the weights 1 to $n_1 + n_2 - 1$ and θ is therefore a graceful valuation of T. Finally if, in the two cases, (i) $\theta_2(u_2) \leq s_2$ when $\theta_1(u_1) > s_1$ or (ii) $\theta_2(u_2) > s_2$ when $\theta_1(u_1) \leq s_1$, then the bases b_1 and b_2 are an even distance

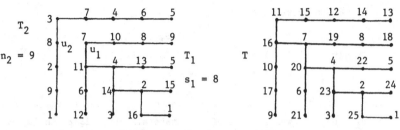

Figure 9 : Thm 8, case (ii).

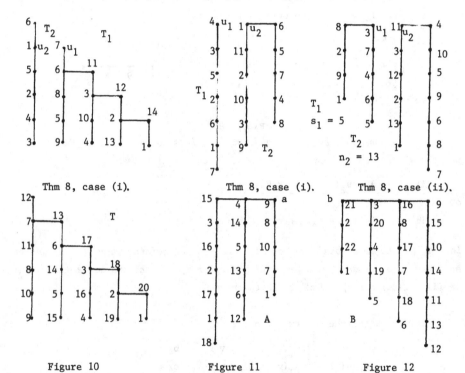

Thm 8, case (i).   Thm 8, case (i).   Thm 8, case (ii).

Figure 10   Figure 11   Figure 12

Thm 8, case (i),
Using A and B.

Figure 13

apart and so, in view of the definition, θ is an interlaced valuation of T with size $s = s_1 + s_2$.

An example of Theorem 7 is shown in Figure (8). Examples derived from Theorem 8 are shown in Figures (9-13). As a starting point for recursive constructions based on these theorems (especially the latter, as in Figures (9-12)) it is interesting to note that for certain n, a chain C_n of vertices in a line has several interlaced valuations: Figures (14,15) are suggestive in this connection; for example, the method of Figure (14) works for all chains C_n with $n \not\equiv 1$ modulo 4. What is the number of interlaced or, for that matter, graceful valuations of C_n?

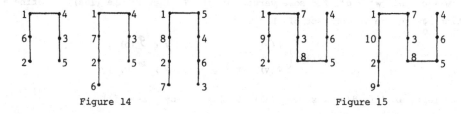

Figure 14   Figure 15

4. AN INTERLACING ALGORITHM

We develop more fully, in this section, an algorithm introduced in [10]. We begin by defining two classes of valuations which prove useful in the ensuing discussion. Firstly, a valuation θ of a tree with a line of symmetry is a *symmetric valuation* if, for symmetrically placed edges (u,v) and (u',v') with $d(u,u') < d(v,v')$ we have either

$$\theta(u) \le \theta(u') < \theta(v') < \theta(v) \tag{4a}$$

or

$$\theta(u) \ge \theta(u') > \theta(v') > \theta(v). \tag{4b}$$

Secondly, an *almost interlaced valuation* of a tree on n vertices, $n > 3$, is a parity valuation under which the weights carried by the $n - 1$ edges take $n - 3$ distinct values, two of these values being repeated.

Let T be a tree on n vertices with interlaced valuation θ having base b and sizes. Let T' be a disjoint copy of the reflection of T, let v' be the image in T' of the vertex v of T and let θ' be the interlaced valuation of T' given by (compare (2))

$$\theta'(v') = s + 1 - \theta(v), \quad \theta(v) \le s \; ; \; = n + s + 1 - \theta(v), \quad s < \theta(v).$$

Likewise, let L be a tree on p vertices, $p \ge 3$, (disjoint from T and T') having an interlaced valuation ξ with base c of valence one and size m. Suppose further that L has a line of symmetry and that, if v' denotes the image of a vertex v in this line, then $m = \xi(c')$ and $d_L(c,c')$ is even. Also let ξ^+ be the interlaced valuation of L given by (compare (1))

$$\xi^+(v) = p + 1 - \xi(v)$$

and let $m^+ = p - m$ (so that m^+ is the size of L under ξ^+).

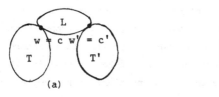 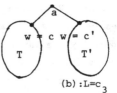

(a) (b) :L=c_3

Figure 16

Now let L * T be the tree obtained from T, T' and L by making the identifications $w = c$ and $w' = c'$ for some vertex w in T (see Figure (16a)). Define a valuation ϕ on the vertex set of L * T by

$$\phi(v) = \begin{cases} \phi(v) & , \quad v \in \mathcal{S}_T(b) , \\ \phi(v) + n + p - 2, & v \notin \mathcal{S}_T(b) ; \end{cases}$$

and, according as (i) $\theta(w) \le s$ or (ii) $\theta(w) > s$, by (case (i))

$$\phi(v) = \begin{cases} \xi(v) + s - 1 & , \quad 1 < \xi(v) < m, \ v \text{ in } L, \\ \xi(v) + n + s - 2, & m < \xi(v), \quad v \text{ in } L , \\ \theta'(v) + s + m - 2, & v \text{ in } T' \end{cases}$$

or (case (ii))

$$\phi(v) = \begin{cases} \xi^+(v) + s & , \quad \xi^+(v) \le m^+, \ v \text{ in } L , \\ \xi^+(v) + n + s - 1, & m^+ + 1 < \xi^+(v) < p, \ v \text{ in } L, \\ \theta'(v) + s + m^+ & , \quad v \text{ in } T'. \end{cases}$$

Then, firstly, L * T has a line of symmetry and ϕ is a symmetric valuation under which, moreover, the weights on symmetrically placed edges (u,v) and (u',v') are related by

$$\omega_\phi(u,v) + \omega_\phi(u',v') = 2n + p - 2.$$

Secondly, ϕ is a parity valuation on L * T such that the weights carried by the edges are all distinct except that the weights on the edges of L incident at $c = w$ and $c' = w'$ may also be carried on edges in T and T' respectively. (Indeed, the edges in T carry all the weights $n + p - 1$ to $2n + p - 3$, those in T' all the weights 1 to $n - 1$ and, for $p > 3$, those in L not incident with $c = w$ or $c' = w'$, all the weights $n + 1$ to $n + p - 3$.) Clearly, if the weights under ϕ are all distinct, then Φ is an interlaced valuation and otherwise it is almost interlaced. (Note that in the latter case only the weights n and $n + p - 2$ are missing.) When can ϕ be modified so as to give an interlaced valuation for L * T?

One case in which we may obtain an interlaced valuation θ of T recursively from ϕ is when L is a chain C_3 on three vertices with endpoints c and c' (see Figure (16b)). In this case, let a be the middle vertex of $L = C_3$, so that a is the apex of L * T. For a tree, such as L * T, with apex a, the *subtree dependent on a vertex* v is the subtree on the vertices u (including v) such that the shortest path from a to u passes through v. Also the endpoints of edges (x,y) of the tree are ordered so that $d(a,x) < d(a,y)$. Our object now is to define inductively a sequence of almost interlaced symmetric valuations ϕ_i of copies $L * T_i$ of L * T which terminates after a finite number of steps, r say, in an interlaced valuation $\theta = \phi_r$ of $L * T_r$ which we identify with L * T.

If ϕ is, to start with, an interlaced valuation of L * T then this objective is already accomplished by taking $\theta = \phi$. Otherwise, suppose that ϕ is not interlaced and let $\phi_1 = \phi$, $T_1 = T$ and $u_0 = u_0' = a$, $v_0 = c$, and $v_0' = c'$. Then ϕ_1 is an almost interlaced, symmetric valuation of $L * T_1$ and there are edges (u_1,v_1) in T_1 and (u_1',v_1') in T_1 such that, for i = 1,

$$\omega_i(u_i,v_i) = \omega_i(u_{i-1},v_{i-1}); \quad \omega_i(u_i',v_i') = \omega_i(u_{i-1}',v_{i-1}') , \quad\quad (5)$$

where $\omega_i(u,v) = |\phi_i(u) - \phi_i(v)|$. (By definition of ϕ, these edges are symmetrically placed, as the notation suggests.) Suppose then that, for $1 \leq i \leq k$, we have obtained almost interlaced symmetric valuations ϕ_i of copies $L * T_i$ of L * T and symmetrically placed edges (u_i,v_i) in T_i (a copy of T) and (u_i',v_i') on T_i such that (5) holds. Let T_{k+1} and T_{k+1}' be the trees obtained from T_k and T_k' by interchanging the subtrees dependent on v_k and v_k' and let ϕ_{k+1} be the valuation (of $L * T_{k+1}$) agreeing with ϕ_k after making this interchange *complete* with the labels on the subtrees under ϕ_k (identifying $L * T_{k+1}$ and $L * T_k$ in the natural way). Then T_{k+1} and $L * T_{k+1}$ are copies of T and L * T respectively and ϕ_{k+1} is also a symmetric parity valuation, as the interchange preserves these properties. However, the previous duplication of weights under ϕ_k has been removed, possibly at the expense of introducing a duplication of the weights under ϕ_{k+1} on the edges (u_k,v_k) and (u_k',v_k'). If there is no such duplication, then ϕ_{k+1} is interlaced and we set $\theta = \phi_{k+1}$. Otherwise, there are symmetrically placed edges (u_{k+1},v_{k+1}) in T_{k+1} and (u_{k+1}',v_{k+1}') in T_{k+1} for which (5) holds with i = k + 1 and we may continue the process. It remains then to show that this inductive procedure terminates with an interlaced valuation.

To this end, we let

$$\delta_i = |\omega_i(u_i,v_i) - \omega_i(u_i',v_i')| , \quad i \leq 1.$$

Then, firstly,

$$\delta_i = |\omega_1(a,c) - \omega_1(a,c')| = |\phi(c) - \phi(c')|$$

which is odd, in view of the symmetry of L * T and the definition of ϕ. Also $\delta_r = 1$ only if $\omega_r(u_r,v_r)$ and $\omega_r(u_r',v_r')$ are the missing weights n + 1 and n, in which case ϕ_r is interlaced. Moreover, for the valuations ϕ_j, $j \geq 1$, obtained above we have, as a result of the interchanges, for $i \geq 2$,

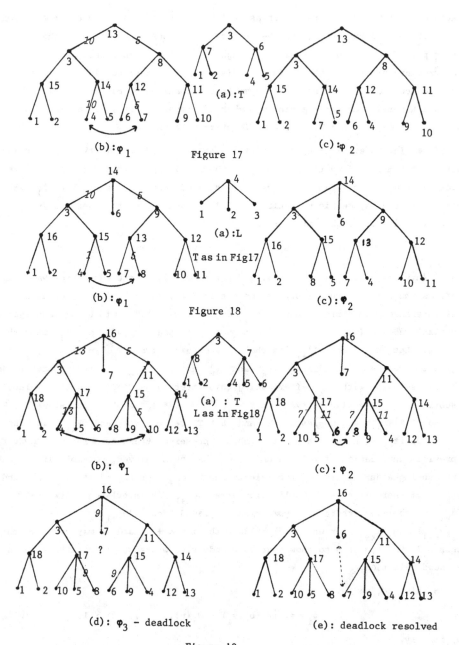

(a):T

(b):φ₁

Figure 17

(c):φ₂

(a):L

T as in Fig17

(b):φ₁

Figure 18

(c):φ₂

(a) : T

L as in Fig18

(b): φ₁

(c): φ₂

(d): φ₃ - deadlock

(e): deadlock resolved

Figure 19

$$\phi_i(u_i) = \phi_{i-1}(u_i) \; ; \; \phi_i(v_i) = \phi_{i-1}(v_i) \; ;$$

$$\phi_i(u_i') = \phi_{i-1}(v_i') \; ; \; \phi_i(v_i') = \phi_{i-1}(u_i') \; ,$$

so, by (5), for $i \geq 1$,

$$\delta_{i+1} = \Big| \, |\phi_i(u_i) - \phi_i(v_i')| - |\phi_i(v_i) - \phi_i(u_i')| \, \Big|.$$

Hence, according as ϕ_i, (u_i, v_i) and (u_i', v_i') satisfy (4a) or (4b) (noting that, by the convention on edges, $d(u_i, u_i') < d(v_i, v_i')$) we have, for $i \geq 1$,

$$\delta_i - \delta_{i+1} = \begin{cases} 2(\phi_i(u_i') - \phi_i(u_i)), & \phi_i(u_i) \leq \phi_i(u_i') < \phi_i(v_i') < \phi_i(v_i) \\ 2(\phi_i(v_i') - \phi_i(v_i)), & \phi_i(v_i) < \phi_i(v_i') < \phi_i(u_i') \leq \phi_i(u_i) \end{cases}.$$

Thus in either case $\delta_i - \delta_{i+1}$ is even and

$$\delta_i > \delta_{i+1} \; , \quad i \geq 1.$$

It follows that the δ_i, $i \geq 1$, are monotonically decreasing odd positive integers and so $\delta_r = 1$ for some $r \geq 1$, in which case $\theta = \phi_r$ is an interlaced valuation of $L * T_r$ and so of $L * T$ as required. It would be of interest to know that value of r for given T (see, for example [10, p.44]). An illustration in which $r = 2$ is shown in Figure (17).

For general L, we may also obtain almost interlaced valuations ϕ_i of $L * T_i$ by the exchange algorithm and in some cases the procedure again terminates in an interlaced valuation (see, for example, Figure (18)). In other cases, however, some sort of deadlock is reached and while, as in Figure (19), we may still sometimes extricate ourselves, it is not clear in general what additional steps need to be incorporated into the algorithm or what further restrictions need to be placed on L and T in order to ensure termination with an interlaced valuation or whether indeed this is always possible. It is also sometimes possible to link up in this way disjoint trees T_i, $1 \leq i \leq m$, each alternately a copy of T or T', but again this needs further detailed investigation.

ACKNOWLEDGEMENTS

We are much obliged to Dr. C.C. Chen, of Nanyang University, for his catalytic agency in bringing about our collaboration. We were unfortunately unable, in preparing this paper, to consult [12] with which it seems some of our work here and in [7,8,9,10] may overlap and of which we only became aware through [5], after much of our work had been completed. Interlaced trees are also useful as building blocks in a construction for graceful trees which we hope to discuss in a further note "Another class of graceful trees" (currently in preparation).

REFERENCES

[1] I. Cahit, 'Are all complete binary trees graceful?', *Amer. Math. Monthly*, 83 (1976), 35 - 37.

[2] C.C. Chen, 'On the enumeration of certain graceful graphs', in *Proceedings of the International Conference on Combinatorial Theory, Canberra, 1977* (to appear).

[3] C.C. Chen and P.Y. Lee, 'Some problems in graph theory', *SEA Bull. Math.*, 1 (1977) 38 - 43; 2 (1978), 39 - 41.

[4] S.W. Golomb, 'How to number a graph', in *Graph Theory and Computing*, ed. R.C. Read, (Academic Press, New York, 1972).

[5] R.K. Guy, 'Monthly research problems', *Amer. Math. Monthly*, 84 (1977), 807 - 815.

[6] K.M. Koh, P.Y. Lee and T. Tan, 'Fibonacci trees', *SEA Bull. Math.*, 2 (1978), 45 - 47.

[7] K.M. Koh, D.G. Rogers and T. Tan, 'On graceful trees', *Nanta Mathematica*, 10 (1977), 207 - 211.

[8] K.M. Koh and T. Tan, 'Two theorems on graceful trees', *Discrete Math.*, (to appear)

[9] K.M. Koh and T. Tan, 'A note on graceful trees', in *Proceedings of the Third Southeast Asian Mathematical Symposium, Universiti Kebangsaan, Malaysia, 1978*, (to appear).

[10] D.G. Rogers, 'A graceful algorithm', *SEA Bull. Math.*, 2 (1978), 42 - 44.

[11] A. Rosa, 'On certain valuations of the vertices of a graph', in *Theory of Graphs, Proceedings of the International Symposium, Rome, 1966*, (Gordon and Breach, New York, 1967).

[12] R.G. Stanton and C.R. Zarnkè, 'Labellings of balanced trees', in *Proceedings of the Fourth South Eastern Conference on Combinatorics, Graph Theory and Computing, Boca Raton 1973, Congressus Numerantium 8*, (Utilitas Mathematica, Winnipeg, 1973).

Department of Mathematics,
Nanyang University,
Singapore.

Mathematical Institute,   and   Department of Mathematics,
Oxford,   University of Western Australia,
England.   Nedlands, 6009,
                                              Western Australia,
                                              Australia.

CONSTRUCTION OF BALANCED DESIGNS AND RELATED IDENTITIES

ELIZABETH J. MORGAN

ABSTRACT

A balanced n-ary design is a design on V elements arranged in B blocks of size K such that each element can occur 0,1,2,..., or n-1 times in each block (so that the blocks are collections of elements rather than subsets) and such that $\sum_{j=1}^{B} n_{ij}n_{mj} = \Lambda$, constant, where n_{ij} is the number of times the ith element occurs in the jth block, $i = 1,...,V$, $j = 1,...,B$. So a balanced binary design is merely a balanced incomplete block design (BIBD).

Any m BIBDs, based on the same set of elements, are used in two constructions to yield balanced (m+1)-ary designs. Some interesting combinatorial identities are involved.

Similar constructions using BIBDs with $\lambda = 1$ and 2 yield other BIBDs.

1. INTRODUCTION

The majority of block designs considered in the literature today consist of blocks which are *subsets* of some set, so that each element (object, variety) may occur at most once in each block. However, as long ago as 1893, E. Hastings Moore [3] considered triple systems in which the triples, or blocks of size 3, were *collections* of elements rather than subsets, in that they could contain each element 0, 1, 2 or 3 times. Moore's one restriction was that each pair of elements should occur in just one block. For example, he gave two possible triple systems on 4 elements:

	aaa	abb	acc	add	bcd ,
and	aaa	abb	acd	bcc	bdd .

In 1952 Tocher [13] introduced a different idea of 'balance' into designs in which blocks are collections of elements. He defined a *balanced n-ary block design* to be a design on V elements with B blocks of size K such that each element occurs $0,1,2,...,$ or n-1 times in each block, and such that

$$\sum_{\substack{j=1 \\ i\neq m}}^{B} n_{ij}n_{mj} = \Lambda \, , \text{ a constant,}$$

where n_{ij} is the number of times the ith element occurs in the jth block, $i = 1,...,V$, $j = 1,...,B$. Thus a balanced binary design is merely a balanced incomplete block design (BIBD). (For definitions and elementary results on BIBDs, the reader is referred to Chapter 7 of [12].)

In this paper we use Tocher's definition and construct balanced n-ary designs from BIBDs. Before doing so we need the following definitions and results.

Let ρ_i, $i = 1,2,\ldots,n-1$, be the number of blocks of a balanced n-ary design in which some element occurs i times. If this number ρ_i is independent of the element chosen, then the replication number R is given by

$$R = \rho_1 + 2\rho_2 + 3\rho_3 + \ldots + (n-1)\rho_{n-1} \ . \tag{1}$$

Such an n-ary design, with constant ρ_i, will be called *regular*. It is straight-forward to verify that the parameters of any regular balanced n-ary design satisfy:

$$VR = BK \ ; \tag{2}$$

$$\Lambda(V-1) = \rho_1(K-1) + 2\rho_2(K-2) + \ldots + (n-1)\rho_{n-1}(K-(n-1)) \ . \tag{3}$$

Note that (3) reduces to the familiar result $\lambda(v-1) = r(k-1)$ in the case of a binary design (BIBD) with parameters (v,b,r,k,λ).

In [4], an analogue of the Bruck-Ryser-Chowla theorem was shown to hold for balanced ternary designs (case n = 3), and some constructions for balanced n-ary designs were given. In the next section here we indicate how the Bruck-Ryser-Chowla theorem holds for any regular balanced n-ary design. In section 3 we give some com-binatorial identities, and use them in section 4 to construct balanced n-ary designs. The methods of construction of the blocks of the designs generalise those given in Morgan [4] and Nigam [7]; this latter paper, [7], has only been brought to the auth-or's attention since completion of this work. Finally in section 5 some BIBDs are constructed along similar lines; this section generalises work in [5].

2. THE BRUCK-RYSER-CHOWLA THEOREM FOR REGULAR BALANCED n-ARY DESIGNS

Theorem 2.1. (Generalised Bruck-Ryser-Chowla.) *In a regular symmetric* (V = B) *balanced n-ary design,*

(i) if V *is even, then* $K^2 - \Lambda V$ *is a perfect square,*
and (ii) if V *is odd, then*

$$z^2 = (K^2 - \Lambda V)x^2 + (-1)^{(V-1)/2}\Lambda y^2$$

has a solution in integers x, y, z *not all zero.* □

We shall not give full details of the proof, which closely follows the proof in the BIBD case (see [12, page 173]). The results follow from the following lemmas.

Lemma 2.2. *Let* N *be the* V × B *incidence matrix* $[n_{ij}]$ *of a regular bal-anced n-ary design. Then*

(i) $NN^T = (KR - \Lambda V)I + \Lambda J$, *where* I *is the* V × V *identity matrix and* J *is the* V × V *matrix with every entry 1;*
and (ii) $\det(NN^T) = KR(KR - \Lambda V)^{V-1}$. □

We omit the proof; see [4] and [12].

<u>Lemma 2.3.</u> *If a regular balanced* n-*ary design satisfies* $KR = \Lambda V$, *then*

(i) *exactly one of* $\rho_1, \rho_2, \ldots, \rho_{n-1}$ *is non-zero,*

(ii) *if* $\rho_i \neq 0$ *then* $K = iV$,

and (iii) *the design consists of* B *complete blocks with each element occurring* i *times.*

<u>Proof.</u> From (1), (3) and $KR = \Lambda V$, we obtain

$$K = \frac{\{R + (2\rho_2 + 6\rho_3 + \ldots + (n-2)(n-1)\rho_{n-1})\}V}{R} . \tag{4}$$

Then from (2) we have

$$B = \frac{R^2}{R + (2\rho_2 + 6\rho_3 + \ldots + (n-2)(n-1)\rho_{n-1})} . \tag{5}$$

The balanced n-ary design clearly must satisfy

$$B \geq \rho_1 + \rho_2 + \ldots + \rho_{n-1} ; \tag{6}$$

however,

$$B - (\rho_1 + \rho_2 + \ldots + \rho_{n-1})$$

$$= \frac{R^2 - (\rho_1 + \ldots + \rho_{n-1})\{R + (2\rho_2 + 6\rho_3 + \ldots + (n-2)(n-1)\rho_{n-1})\}}{R + (2\rho_2 + 6\rho_3 + \ldots + (n-2)(n-1)\rho_{n-1})}$$

$$= \frac{-\sum_{1 \leq i < j \leq n-1} (j-i)^2 \rho_i \rho_j}{R + (2\rho_2 + 6\rho_3 + \ldots + (n-2)(n-1)\rho_{n-1})} \leq 0 ,$$

with equality here if and only if precisely one of the ρ_i is non-zero. The results now follow. □

Henceforth we assume that our n-ary designs are *not* trivially complete in the sense of the above lemma, so that $KR \neq \Lambda V$.

<u>Lemma 2.4.</u> (i) *In a regular balanced* n-*ary design,* $B \geq V$.

(ii) *In a regular symmetric balanced* n-*ary design with incidence matrix* N,

$$\det N = K(K^2 - \Lambda V)^{(V-1)/2}. \qquad\qquad □$$

The above lemma follows directly from Lemmas 2.2 and 2.3. The proof of Theorem 2.1 now proceeds in the same way as the binary case proof.

3. SOME COMBINATORIAL IDENTITIES

In section 4 we use m BIBDs, all based on a common v-set, to construct balanced (m+1)-ary designs. We let the parameters of these BIBDs be $(v, b_i, r_i, k_i, \lambda_i)$, $1 \leq i \leq m$. Consequently we state the results in this section in terms of such parameters, ready for subsequent use, although at no stage do we use the relationships

$vr_i = b_i k_i$ or $\lambda_i(v-1) = r_i(k_i-1)$ between these parameters.

We let the symbol $\sum\limits_{m}^{*}$ in expressions such as

$$\sum_{m}^{*} r_{i_1} r_{i_2} \cdots r_{i_s} (b_{i_{s+1}} - r_{i_{s+1}}) \cdots (b_{i_m} - r_{i_m})$$

mean the sum, for fixed s, $0 \le s \le m$, of all the possible combinations, where $\{i_1, i_2, \ldots, i_s, i_{s+1}, \ldots, i_m\} = \{1, 2, \ldots, m\}$, with $\binom{m}{s}$ terms in the sum.

<u>Lemma 3.1.</u> *Let* $X = \sum\limits_{m}^{*} r_{i_1} \cdots r_{i_s} (b_{i_{s+1}} - r_{i_{s+1}}) \cdots (b_{i_m} - r_{i_m})$. *Then*

(i) $\displaystyle\sum_{s=0}^{m} X = b_1 b_2 \cdots b_m$;

(ii) $\displaystyle\sum_{s=0}^{m} sX = \sum_{i=1}^{m} \frac{r_i}{b_i} b_1 b_2 \cdots b_m$;

(iii) $\displaystyle\sum_{s=0}^{m} s(s-1)X = 2 \sum_{1 \le i < j \le m} \frac{r_i r_j}{b_i b_j} b_1 b_2 \cdots b_m$;

(iv) $\displaystyle\sum_{s=0}^{m} s^2 X = \sum_{i=1}^{m} \frac{r_i}{b_i} b_1 b_2 \cdots b_m + 2 \sum_{1 \le i < j \le m} \frac{r_i r_j}{b_i b_j} b_1 b_2 \cdots b_m$.

<u>Proof.</u> Consider the expression

$$\prod_{i=1}^{m} (xr_i + (b_i - r_i)) \ . \tag{7}$$

When expanded, (7) is a sum of terms of the form

$$x^s r_{i_1} r_{i_2} \cdots r_{i_s} (b_{i_{s+1}} - r_{i_{s+1}}) \cdots (b_{i_m} - r_{i_m}) \ .$$

Thus (7) equals

$$\sum_{s=0}^{m} (x^s X) \ . \tag{8}$$

Differentiating (7) and (8) t times with respect to x, and then letting $x = 1$, for $t = 0, 1, 2$ yields *(i)*, *(ii)* and *(iii)* respectively; *(iv)* then follows. □

Note also the identity

$$\prod_{i=1}^{m} [x\lambda_i + y(r_i - \lambda_i) + z(r_i - \lambda_i) + (b_i - 2r_i + \lambda_i)] =$$

$$\sum_{a+b+c+d=m} \left\{ \sum_{m}^{*} x^a y^b z^c \lambda_{i_1} \cdots \lambda_{i_a} (r_{j_1} - \lambda_{j_1}) \cdots (r_{j_b} - \lambda_{j_b})(r_{k_1} - \lambda_{k_1}) \cdots (r_{k_c} - \lambda_{k_c}) \times \right.$$

$$\left. (b_{\ell_1} - 2r_{\ell_1} + \lambda_{\ell_1}) \cdots (b_{\ell_d} - 2r_{\ell_d} + \lambda_{\ell_d}) \right\} \ . \tag{9}$$

Differentiating both sides of (9) with respect to x, y and/or z where appropriate, and letting $x = y = z = 1$, yields the following results:

Lemma 3.2. *Let*

$$X = \sum_{m}^{*} \lambda_{i_1}\cdots\lambda_{i_a}(r_{j_1}-\lambda_{j_1})\cdots(r_{j_b}-\lambda_{j_b})(r_{k_1}-\lambda_{k_1})\cdots(r_{k_c}-\lambda_{k_c})(b_{\ell_1}-2r_{\ell_1}+\lambda_{\ell_1})\cdots$$
$$(b_{\ell_d}-2r_{\ell_d}+\lambda_{\ell_d}) ,$$

where $a+b+c+d = m$, *and* $\{i_1,\ldots,i_a,j_1,\ldots,j_b,k_1,\ldots,k_c,\ell_1,\ldots,\ell_d\} = \{1,2,\ldots,m\}$ *in some order. Then*

(i) $\displaystyle\sum_{a=0}^{m}\sum_{b=0}^{m-a}\sum_{c=0}^{m-a-b} X = b_1 b_2 \cdots b_m$;

(ii) $\displaystyle\sum_{a=0}^{m}\sum_{b=0}^{m-a}\sum_{c=0}^{m-a-b} aX = \sum_{i=1}^{m}\frac{\lambda_i}{b_i} b_1 b_2 \cdots b_m$;

(iii) $\displaystyle\sum_{a=0}^{m}\sum_{b=0}^{m-a}\sum_{c=0}^{m-a-b} a^2X = \sum_{i=1}^{m}\frac{\lambda_i}{b_i} b_1\cdots b_m + 2\sum_{1\le i<j\le m}\frac{\lambda_i\lambda_j}{b_i b_j} b_1\cdots b_m$;

(iv) $\displaystyle\sum_{a=0}^{m}\sum_{b=0}^{m-a}\sum_{c=0}^{m-a-b} abX = \sum_{1\le i<j\le m}\frac{(r_i-\lambda_i)\lambda_j + \lambda_i(r_j-\lambda_j)}{b_i b_j} b_1 b_2 \cdots b_m$;

(v) $\displaystyle\sum_{a=0}^{m}\sum_{b=0}^{m-a}\sum_{c=0}^{m-a-b} bcX = 2\sum_{1\le i<j\le m}\frac{(r_i-\lambda_i)(r_j-\lambda_j)}{b_i b_j} b_1 b_2 \cdots b_m$;

(vi) $\displaystyle\sum_{a=0}^{m}\sum_{b=0}^{m-a}\sum_{c=0}^{m-a-b} (a+b)(a+c)X = \left\{\sum_{i=1}^{m}\frac{\lambda_i}{b_i} + 2\sum_{1\le i<j\le m}\frac{r_i r_j}{b_i b_j}\right\} b_1 b_2 \cdots b_m$. $\quad\square$

4. CONSTRUCTION OF BALANCED n-ARY DESIGNS

Since Tocher's paper [13] appeared in 1952, several constructions of balanced n-ary designs have appeared; see [2, 4, 6, 7, 8, 9, 10, 11]. Various constructions depend on mutually orthogonal Latin squares, difference sets from finite fields, and incidence matrices of BIBDs (adding rows and columns in certain ways).

Henceforth let us suppose that we have m BIBDs all based on the same v-set of elements, with parameters $(v,b_i,r_i,k_i,\lambda_i)$, $1 \le i \le m$. We shall construct regular balanced $(m+1)$-ary designs in two ways; in each case the $(m+1)$-ary design is based on the same v-set as the m BIBDs so that $V = v$. Also the blocks of the $(m+1)$-ary design are obtained by taking the strong union (with repetitions of elements retained) of certain blocks of the BIBDs.

Construction I. The blocks of the $(m+1)$-ary design are formed by taking the strong union of m blocks, one from each of the m BIBDs, in all possible ways. Thus B new blocks are formed where $B = b_1 b_2 \cdots b_m$, and each is of size K where $K = k_1 + k_2 + \ldots + k_m$.

Construction II. Each block of the $(m+1)$-ary design is formed by taking the strong union of m blocks all containing a common element x, one chosen from each of the m BIBDs. This is done in all possible ways, and for each of the v elements x, so that B new blocks are formed where $B = vr_1 r_2 \ldots r_m$; each is of size K where $K = k_1 + k_2 + \ldots + k_m$. We shall refer to such new blocks of size K formed from m BIBD blocks each containing element x as "new x-blocks".

We shall now verify that each of these constructions does indeed yield a regular balanced $(m+1)$-ary design; clearly we only need to check the parameters ρ_i, R and Λ.

From the block description in construction I, we have, for $1 \le s \le m$,

$$\rho_s = \sum_m^* r_{i_1} r_{i_2} \ldots r_{i_s} (b_{i_{s+1}} - r_{i_{s+1}}) \ldots (b_{i_m} - r_{i_m}) . \tag{10}$$

Now from (1), (10) and Lemma 3.1 (ii) we obtain

$$R = \sum_{i=1}^m \frac{r_i}{b_i} b_1 b_2 \ldots b_m = \frac{BK}{V} \quad \text{as required.}$$

To verify the balance, let x and y be an arbitrary pair of elements. Let Λ_{xy} denote $\sum_{j=1}^B n_{x_j} n_{y_j}$ where n_{x_j}, n_{y_j} are respectively the number of occurrences of x and y in block j. Suppose that among the m blocks chosen from the BIBDs there are a blocks which contain both x and y, b which contain x but not y, c which contain y but not x, and $d = m-a-b-c$ which contain neither x nor y. Then the new block formed from the strong union of these m blocks will contribute $(a+b)(a+c)$ to Λ_{xy}. Altogether,

$$\Lambda_{xy} = \sum_{a=0}^m \sum_{b=0}^{m-a} \sum_{c=0}^{m-a-b} (a+b)(a+c) \left\{ \sum_m^* \lambda_{i_1} \ldots \lambda_{i_a} (r_{j_1} - \lambda_{j_1}) \ldots (r_{j_b} - \lambda_{j_b}) \times \right.$$
$$\left. (r_{k_1} - \lambda_{k_1}) \ldots (r_{k_c} - \lambda_{k_c}) (b_{\ell_1} - 2r_{\ell_1} + \lambda_{\ell_1}) \ldots (b_{\ell_d} - 2r_{\ell_d} + \lambda_{\ell_d}) \right\} .$$

From Lemma 3.2 (vi), this equals

$$\left\{ \sum_{i=1}^m \frac{\lambda_i}{b_i} + 2 \sum_{1 \le i < j \le m} \frac{r_i r_j}{b_i b_j} \right\} b_1 b_2 \ldots b_m ,$$

and is clearly independent of our choice of x and y, so let $\Lambda_{xy} = \Lambda$. It can also be verified that this value of Λ agrees with (3); details are omitted, but (ii) and (iv) of Lemma 3.1 are used in this verification. Thus we have

Theorem 4.1. *If there exist m BIBDs with parameters $(v, b_i, r_i, k_i, \lambda_i)$, $i = 1, 2, \ldots, m$, then there exists a regular balanced $(m+1)$-ary design with parameters*

$$V = v, \quad B = b_1 b_2 \ldots b_m, \quad K = k_1 + k_2 + \ldots + k_m,$$

$$\rho_s = \sum_m^* r_{i_1} \ldots r_{i_s} (b_{i_{s+1}} - r_{i_{s+1}}) \ldots (b_{i_m} - r_{i_m}), \quad 1 \le s \le m,$$

$$R = \sum_{i=1}^m \frac{r_i}{b_i} b_1 \ldots b_m, \quad \Lambda = \sum_{i=1}^m \frac{\lambda_i}{b_i} b_1 \ldots b_m + 2 \sum_{1 \le i < j \le m} \frac{r_i r_j}{b_i b_j} b_1 \ldots b_m . \qquad \Box$$

Now consider construction II. Let x be any one of the v elements. There are $r_1 r_2 \ldots r_m$ new x-blocks, which each contain m copies of x, and there are $(v-1)\lambda_1 \lambda_2 \ldots \lambda_m$ other new blocks also each containing m copies of x. Thus

$$\rho_m = (v-1)\lambda_1 \lambda_2 \ldots \lambda_m + r_1 r_2 \ldots r_m . \tag{11}$$

Let z be any element distinct from x. Consideration of occurrences of x in the new z-blocks shows that

$$\rho_s = (v-1) \sum_m^* \lambda_{i_1} \ldots \lambda_{i_s} (r_{i_{s+1}} - \lambda_{i_{s+1}}) \ldots (r_{i_m} - \lambda_{i_m}) , \quad 1 \le s \le m-1. \tag{12}$$

Hence we have

$$R = r_1 r_2 \ldots r_m \left\{ (v-1) \sum_{i=1}^m \frac{\lambda_i}{r_i} + m \right\} ;$$

this follows from (1), (11), (12), and Lemma 3.1 *(ii)* with r_i, b_i there changed to λ_i, r_i respectively. Thus

$$R = r_1 r_2 \ldots r_m (k_1 + k_2 + \ldots + k_m) = \frac{BK}{V} , \quad \text{as required.}$$

To verify the balance of the (m+1)-ary design, as in construction I we calculate Λ_{xy} for an arbitrary pair of elements x and y. New x-blocks each contain m copies of x and some number of copies of y; if there are s copies of y in a new x-block it contributes sm to Λ_{xy}. Thus all the new x-blocks contribute

$$\sum_{s=1}^m sm \sum_m^* \lambda_{i_1} \ldots \lambda_{i_s} (r_{i_{s+1}} - \lambda_{i_{s+1}}) \ldots (r_{i_m} - \lambda_{i_m})$$

to Λ_{xy}, as do the new y-blocks.

Now let z be any element distinct from x and y. Suppose some new z-block consists of m BIBD blocks, where a of them contain z, x and y, b contain z, x but not y, c contain z, y but not x, and $d = m-a-b-c$ contain z but not x or y. Such a new z-block contributes $(a+b)(a+c)$ to Λ_{xy}. Let w_i^z be the number of blocks of the ith BIBD, $1 \le i \le m$, which contain z, x and y; this number w_i^z is a function of z, x and y, and is *not* constant unless the ith BIBD is a t-design with $t \ge 3$. However,

$$\sum_{z \ne x,y} w_i^z = \lambda_i (k_i - 2) , \quad 1 \le i \le m , \tag{13}$$

where the sum here is over all $v-2$ elements z distinct from x and y. For some fixed element $z \ne x,y$, all the new z-blocks give a contribution to Λ_{xy} of

$$\sum_{a=0}^m \sum_{b=0}^{m-a} \sum_{c=0}^{m-a-b} (a+b)(a+c) \left\{ \sum_m^* w_{i_1}^z \ldots w_{i_a}^z (\lambda_{j_1} - w_{j_1}^z) \ldots (\lambda_{j_b} - w_{j_b}^z)(\lambda_{k_1} - w_{k_1}^z) \ldots \right.$$

$$\left. (\lambda_{k_c} - w_{k_c}^z)(r_{\ell_1} - 2\lambda_{\ell_1} + w_{\ell_1}^z) \ldots (r_{\ell_d} - 2\lambda_{\ell_d} + w_{\ell_d}^z) \right\} . \tag{14}$$

If we now use Lemma 3.2 *(vi)* with λ, r and b there replaced by w^z, λ and r

respectively, then (14) equals

$$r_1 r_2 \dots r_m \left\{ \sum_{i=1}^{m} \frac{w_i^z}{r_i} + 2 \sum_{1 \le i < j \le m} \frac{\lambda_i \lambda_j}{r_i r_j} \right\}.$$

Now summing for all z distinct from x and y, using (13), and including the contributions from new x-blocks and new y-blocks, we have

$$\Lambda_{xy} = 2m \sum_{s=1}^{m} s \left\{ \sum_{m}^{*} \lambda_{i_1} \dots \lambda_{i_s} (r_{i_{s+1}} - \lambda_{i_{s+1}}) \dots (r_{i_m} - \lambda_{i_m}) \right\}$$
$$+ \sum_{i=1}^{m} \frac{\lambda_i (k_i - 2)}{r_i} r_1 \dots r_m + 2(v-2) \sum_{1 \le i < j \le m} \frac{\lambda_i \lambda_j}{r_i r_j} r_1 \dots r_m.$$

This is clearly independent of x and y, and by means of Lemma 3.1 appropriately amended, we obtain

$$\Lambda_{xy} = \Lambda = \left\{ \sum_{i=1}^{m} \frac{\lambda_i}{r_i} r_1 \dots r_m \right\} (k_1 + \dots + k_m + m - 1) - 2 \sum_{1 \le i < j \le m} \frac{\lambda_i \lambda_j}{r_i r_j} r_1 \dots r_m.$$

Once again it can be shown that this agrees with the value of Λ obtained from (3). Thus we have

Theorem 4.2. *If there exist m BIBDs with parameters $(v, b_i, r_i, k_i, \lambda_i)$, $1 \le i \le m$, then there exists a regular balanced (m+1)-ary design with parameters*

$$V = v, \quad B = v r_1 r_2 \dots r_m, \quad K = k_1 + k_2 + \dots + k_m,$$

$$\rho_s = (v-1) \sum_{m}^{*} \lambda_{i_1} \dots \lambda_{i_s} (r_{i_{s+1}} - \lambda_{i_{s+1}}) \dots (r_{i_m} - \lambda_{i_m}), \quad 1 \le s \le m-1,$$

$$\rho_m = (v-1) \lambda_1 \lambda_2 \dots \lambda_m + r_1 r_2 \dots r_m, \quad R = r_1 r_2 \dots r_m (k_1 + k_2 + \dots + k_m),$$

$$\Lambda = r_1 r_2 \dots r_m \left\{ (k_1 + \dots + k_m + m - 1) \sum_{i=1}^{m} \frac{\lambda_i}{r_i} - 2 \sum_{1 \le i < j \le m} \frac{\lambda_i \lambda_j}{r_i r_j} \right\}. \qquad \square$$

Remarks and Examples.

Consider the following two representations of the (7,7,3,3,1)-design, both based on the set $\{1,2,3,4,5,6,7\}$.

D_1: 123, 145, 167, 246, 257, 347, 356.

D_2: 124, 235, 346, 457, 561, 672, 713.

Construction I with $m = 2$ yields a regular balanced ternary design (regular BTD) with parameters

$$(V, B; \rho_1, \rho_2, R; K, \Lambda) = (7, 49; 24, 9, 42; 6, 32),$$

while construction II with $m = 2$ yields a regular BTD with parameters (7, 63; 24, 15, 54; 6, 40). It is interesting to note that in each construction, different representations of the same (that is, isomorphic) BIBD may result in non-isomorphic n-ary designs with necessarily the same parameters. For instance, if two identical copies of design D_1 above are used in construction I, the resulting BTD contains 7 blocks of the form aabbcc, where abc is a block of D_1, whereas in the same construction using designs D_1 and D_2, clearly no block of the form aabbcc can result, since there is no block abc common to the two representations D_1 and D_2. Similarly, in construction II, use of two identical copies of D_1 results in a BTD containing 7 different blocks of type aabbcc, each occurring 3 times, as a new a-block, new b-block and new c-block. But no such new blocks arise from construction II if representations D_1 and D_2 are used.

Generally, if m identical copies of one BIBD with parameters (v,b,r,k,λ) are used in the constructions above, certain "repetitious" blocks may be omitted without affecting the balance of the (m+1)-ary design. In construction I, if C is a block of the BIBD, there will be b blocks of the form CC...C (m copies) in the (m+1)-ary design, one for each block C in the BIBD as C varies; deletion of these b blocks will reduce B by b, R by mr, ρ_m by r and Λ by $m^2\lambda$, while V and K will be unchanged. In construction II, if C is any BIBD block there will be k identical copies of block CC...C (m copies), and thus bk blocks of this form altogether, as C varies. Deletion of these bk blocks will not affect the balance of the (m+1)-ary design; it will reduce B by bk, R by mrk, ρ_m by rk and Λ by $m^2 k\lambda$.

5. CONSTRUCTION OF BIBDs

For constructions of the type considered above to yield BIBDs, the first obvious requirement is that ordinary union rather than strong union of blocks should be taken. Secondly, when repeats of elements are discarded from n-ary design blocks, the resulting design should have constant block size. However this is not sufficient to ensure balance, as the following example shows. Tocher [13], in the appendix, gives the following regular symmetric BTD with parameters

$$V = B = 6, \quad R = K = 4, \quad \rho_1 = 2, \quad \rho_2 = 1, \quad \Lambda = 2.$$

1146, 2215, 3315, 4423, 5546, 6623.

Here the deletion of repeats does not produce a balanced design.

Let construction I be applied in the case m = 2 to two identical copies of a symmetric BIBD (SBIBD) with parameters (v,v,k,k,λ). When blocks of the form $x_1 x_1 x_2 x_2 \ldots x_k x_k$ are discarded, and when blocks B_i and B_j are paired together only once, so that $\binom{v}{2}$ blocks are formed, ordinary union instead of strong union produces a BIBD, as was shown in [5, Theorem 4]; it has parameters

$$\left(v, \binom{v}{2}, \frac{(2k-\lambda)(v-1)}{2}, 2k-\lambda, \binom{2k-\lambda}{2} \right). \tag{15}$$

This will not generalise to the situation in which m identical copies of an SBIBD are used, with $m \geq 3$, because the size of the intersection of three or more blocks of an SBIBD is not known in general. However a form of construction II in the case $\lambda = 1$ will work for arbitrary m. Let us take the ordinary union of m distinct blocks of a BIBD with parameters $(v,b,r,k,1)$, all of which contain a common

element x.   Doing this in all possible ways, and for each of the v elements x,
results in a BIBD; we omit details but record the result:

Theorem 5.1.   *If there exists a BIBD with parameters* $(v,b,r,k,1)$, *then there
exists a BIBD with parameters*

$$\left(v,\ v\binom{r}{m},\ \binom{r}{m}(mk-m+1),\ mk-m+1,\ \binom{r-1}{m-1}(mk-m+1)\right),$$

for all m *such that* $2 \leq m \leq r-1$.   □

Note that m = 1 is excluded here because in that case the resulting design
consists of k copies of the original design.   Also m = r is excluded because
then the resulting design is complete, with v blocks of size v.

The above theorem generalises Theorems 1 and 2 of [5].

The following example shows that $\lambda = 1$ is a necessary condition in Theorem 5.1
if the original BIBD is not symmetric.   Let the blocks of the (6,10,5,3,2)-design be
      123,   124,   135,   146,   156,   236,   245,   256,   345,   346.
Then blocks formed from the (ordinary) union of two blocks containing element 1 include 1234 and 12356, so the block size in the resulting design is not even
constant.

The next example shows that if λ is at least 3, even using m identical copies
of a *symmetric* design in this adapted form of construction II will not suffice, if
m > 2.   Consider the SBIBD with parameters (25,25,9,9,3) listed in the appendix of
Hall [1] (where the third block in the right hand list there should read
$p\ i\ g\ o\ t\ s\ w\ x\ h$).   Five of the 9 blocks of this design containing element a are

$$a\ b\ c\ d\ e\ f\ g\ h\ i$$
$$q\ a\ b\ p\ u\ c\ x\ y\ t$$
$$r\ c\ s\ a\ o\ w\ b\ v\ n$$
$$g\ r\ m\ l\ d\ a\ q\ s\ p$$
$$m\ o\ f\ n\ y\ l\ h\ a\ x\ .$$

The union of the first three blocks above contains 21 elements, while the union of
the first block above with the last two contains only 19 elements.   So again block
size is not constant.

In the case of a symmetric design with $\lambda = 2$, our construction of type II
works, using m identical copies of a (v,v,k,k,2)-design.   Thus we have

Theorem 5.2.   *If there exists an SBIBD with parameters* $(v,v,k,k,2)$,   *then
there exists a BIBD with parameters*

$$\left(v,\ v\binom{k}{m},\ \binom{k}{m}(km+1-\tfrac{1}{2}m(m+1)),\ km+1-\tfrac{1}{2}m(m+1),\ (km+1-\tfrac{1}{2}m(m+1))\left[2\binom{k-1}{m-1}-\binom{k-2}{m-2}\right]\right),$$

for all m *such that* $2 \leq m \leq k-2$.

Proof.   Bearing in mind that new blocks are formed from the union of m distinct blocks containing an element x,   chosen in all possible ways and for each

element x, we rule out the cases m = 1, k-1 and k for the following reasons.
When m = 1, we obtain k identical copies of our original SBIBD.   For incomplete-
ness we require $km + 1 - \frac{1}{2}m(m+1) < v$.   Since $2(v-1) = k(k-1)$, this is equivalent
to (m-k)(m-k+1) > 0, and so cases m = k and k-1 yield complete designs. Note
also that when m = 2 we obtain two copies of the design with parameters (15).

Clearly the construction yields a design on v elements with $v\binom{k}{m}$ blocks.
To verify constant block size we use induction on m; if m = 2, the union of any
two SBIBD blocks is of size 2k-2 since an SBIBD is a linked design.   Now suppose
that the union of any m blocks containing a common element x is of size
$km + 1 - \frac{1}{2}m(m+1)$.   Let us adjoin another SBIBD block containing x to this union.
Besides x, this new block will have one element in common with each of the m
blocks (from the linkage of the SBIBD), and so only k-m-1 new elements are adjoined.
Thus the union of m+1 blocks containing x is of size
$$km + 1 - \tfrac{1}{2}m(m+1) + (k-m-1) = k(m+1) + 1 - \tfrac{1}{2}(m+1)(m+2) ,$$
as required.

As before, let us use the terms "old x-block" and "new x-block" respectively,
when referring to a block of the SBIBD containing element x, or to a block of the
new design formed from m SBIBD blocks containing x.   Let us count occurrences of
an element x in new blocks, in order to verify constant replication in the new
design.   Each of the $\binom{k}{m}$ new x-blocks contains x.   Let y be any element distinct from x; a new y-block containing x arises either from taking both old blocks
containing x and y, and m-2 more old y-blocks, or else from taking one of the
two old blocks containing both x and y, and another m-1 containing y but not
x.   Thus x occurs in
$$\binom{k}{m} + (v-1)\left\{\binom{k-2}{m-2} + 2\binom{k-2}{m-1}\right\} = \binom{k}{m}(km + 1 - \tfrac{1}{2}m(m+1))$$
new blocks altogether.

In order to check balance, let x and y be an arbitrary pair of elements,
occurring in λ_{xy} new blocks, and suppose that B_1 and B_2 are the two blocks of
the SBIBD containing both x and y.   A simple count shows that the new x-blocks
contribute $\binom{k-2}{m-2} + 2\binom{k-2}{m-1}$ to λ_{xy}, as do the new y-blocks.   Each element distinct
from x and y is one of two possible types: there are 2(k-2) elements z which
occur in one or other of B_1 and B_2, and there are v-2k+2 elements w which do
not.   Considering a 'z-type' element first, there is one old block containing z, x
and y, one containing z and x but not y, one containing z and y but not
x, and k-3 containing z but not x or y.   Thus the new z-blocks (for each z)
contribute $\binom{k-3}{m-3} + 3\binom{k-3}{m-2} + \binom{k-3}{m-1}$ to λ_{xy}.

Now consider a 'w-type' element.   There are two old blocks containing w and
x but not y, two containing w and y but not x, and so k-4 containing w

but not x or y. Thus the new w-blocks, for each w, contribute

$$\binom{k-4}{m-4} + 4\binom{k-4}{m-3} + 4\binom{k-4}{m-2}$$

to λ_{xy}. Therefore

$$\lambda_{xy} = 2\binom{k-2}{m-2} + 4\binom{k-2}{m-1} + 2(k-2)\left\{\binom{k-3}{m-3} + 3\binom{k-3}{m-2} + \binom{k-3}{m-1}\right\}$$

$$+ (v - 2k + 2)\left\{\binom{k-4}{m-4} + 4\binom{k-4}{m-3} + 4\binom{k-4}{m-2}\right\}.$$

This can be shown to simplify to the expression for λ given in the statement of the theorem. □

Remark. There are no small examples to illustrate Theorem 5.2. The requirement $v \geq 2k$, so that the complementary design does not have smaller blocks, translates to $k^2 - k(4m+1) + 2(m^2+m-1) \geq 0$ here. It follows that we need $2k \leq 4m+1 - \sqrt{(8m^2+9)}$ or $2k \geq 4m+1 + \sqrt{(8m^2+9)}$. However since we also require $k \geq m+2$, the smaller possibility for k cannot arise. For example, if $m = 3$, we must take $k \geq 11$, and if $m = 4$, then $k \geq 15$, for the resulting BIBD to have blocks of size less than or equal to $\frac{1}{2}v$.

6. ACKNOWLEDGEMENTS

This work forms part of one chapter of my Ph.D. thesis; I wish to thank my supervisor Dr. Anne Penfold Street for her help. Also I thank Dr. A.S. Jones who suggested the method of proof for Lemmas 3.1 and 3.2, after I had used the much longer method of induction on m.

REFERENCES

[1] Marshall Hall Jr., *Combinatorial Theory*, Blaisdell, Waltham, Mass., 1967.

[2] S.K. Mehta, S.K. Agarwal and A.K. Nigam, On partially balanced incomplete block designs through partially balanced ternary designs, *Sankhya* 37 (B) (1975), 211-219.

[3] E. Hastings Moore, Concerning triple systems, *Math. Ann.* 43 (1893), 271-285.

[4] Elizabeth J. Morgan, Construction of balanced n-ary designs, *Utilitas Mathematica* 11 (1977), 3-31.

[5] Elizabeth J. Morgan, Construction of balanced incomplete block designs, *J. Aust. Math. Soc.* 23(A) (1977), 348-353.

[6] J.S. Murty and M.N. Das, Balanced n-ary block designs and their uses, *Jour. Ind. Stat. Assoc.* 5 (1968), 73-82.

[7] A.K. Nigam, On construction of balanced n-ary block designs and partially balanced arrays, *J. Ind. Soc. Agri. Stat.* 26 (1974), 48-56.

[8] A.K. Nigam, S.K. Mehta and S.K. Agarwal, Balanced and nearly balanced n-ary designs with varying block sizes and replications, *J. Ind. Soc. Agri. Stat.* 30 (1977), 92-96.

[9] G.M. Saha and A. Dey, On construction and uses of balanced n-ary designs, *Ann. Inst. Statist. Math., Tokyo Ann.* 25 (1973), 439-445.

[10] G.M. Saha, On construction of balanced ternary designs, *Sankhya* 37 (B) (1975), 220-227.

[11] S.D. Sharma and B.L. Agarwal, Some aspects of construction of balanced n-ary designs, *Sankhya* 38 (B) (1976), 199-201.

[12] Anne Penfold Street and W.D. Wallis, *Combinatorial Theory: An Introduction,* The Charles Babbage Research Centre, Winnipeg, Canada, 1977.

[13] K.D. Tocher, The design and analysis of block experiments, *Jour. Roy. Statist. Soc.* B 14 (1952), 45-100.

Department of Mathematics
University of Queensland
St. Lucia
Queensland

A GENERALIZATION OF A COVERING PROBLEM OF
MULLIN AND STANTON FOR MATROIDS

James G. Oxley

A (q,m,k) *cover of* V(k,q), *the vector space of k-tuples over* GF(q), *is a sub-set of the non-zero vectors of* V(k,q) *which has rank k and has non-empty intersection with every subspace of* V(k,q) *of rank k - m. A* (q,m,k) *cover may also be viewed as a matroid. As such it is essentially the image in* V(k,q) *of a restriction M of* PG(k-1,q) *under some representation, where M has rank k and critical exponent greater than m. An earlier paper answered several questions of Mullin and Stanton concerning* (2,m,k) *covers. This paper answers the corresponding questions for* (q,m,k) *covers when q > 2. In particular, the least number* η(q,m,k) *of elements in a* (q,m,k) *cover is determined and those matroids having exactly* η(q,m,k) *elements are characterized.*

INTRODUCTION

This paper is concerned with the critical problem for matroids and is essentially an addendum to an earlier paper [8]. We show that with only slight modifications the arguments of [8] can be extended from binary matroids to matroids representable over an arbitrary finite field.

The terminology used here for matroids and graphs will in general follow Welsh [9] and Bondy and Murty [1] respectively. If M is a matroid on a set S and $A \subseteq S$, then rk A and cork A will denote respectively the rank and corank of A in M. The rank and corank of M will also be denoted by rk M and cork M respectively. We shall sometimes denote the restriction of M to S\T by M\T or, if $T = \{x_1, x_2, \ldots, x_n\}$, by $M \backslash x_1, x_2, \ldots, x_n$. Likewise the contraction of M to S\T will sometimes be written as M\T or $M \backslash x_1, x_2, \ldots, x_n$. The set of positive integers will be denoted by $\mathbb{Z}^+$.

If A is a subset of V(n,q), the vector space of n-tuples over GF(q), then a j-tuple $(f_1, f_2, \ldots, f_j)$ of linear functionals on V(n,q) is said to *distinguish* A if for all e in A, we have $f_i(e) \neq 0$ for some i in $\{1, 2, \ldots, j\}$. Let M be a rank n matroid on a set S. The *chromatic polynomial* $P(M; \lambda)$ of M is defined by

$$P(M; \lambda) = \sum_{A \subseteq S} (-1)^{|A|} \lambda^{n-\text{rk } A}.$$

Now suppose that M is representable over GF(q) and let ϕ be a representation of M in V(n,q). Then if $j \in \mathbb{Z}^+$, by [4, p.16.4], the number of j-tuples of linear functionals on V(n,q) which distinguish $\phi(S)$ equals $P(M; q^j)$. Thus $P(M; q^j) \geq 0$ for all j in $\mathbb{Z}^+$. The *critical exponent* c(M; q) of M is defined by

$$c(M; q) = \begin{cases} \infty & \text{, if M has a loop;} \\ \min\{j \in \mathbb{Z}^+ : P(M, q^j) > 0\}, & \text{otherwise.} \end{cases}$$

The critical exponent has the following alternative interpretation. If M is a rank n, loopless matroid representable over GF(q) and ϕ is a representation of M in $V(n,q)$, then $c(M; q)$ is the least number j of hyperplanes $H_1, H_2, \ldots, H_j$ of $V(n,q)$ such that

$$(\bigcap_{i=1}^{j} H_i) \cap \phi(S) = \emptyset.$$

If $k, m \in \mathbb{Z}^+$ and $1 \le m \le k - 1$, then a (q,m,k) *cover of* $V(k,q)$ is a subset of the non-zero vectors of $V(k,q)$ which has rank k and has non-empty intersection with every subspace of $V(k,q)$ of rank $k - m$. A (q,m,k) cover is *simple* if it contains at most one member of each one-dimensional subspace of $V(k,q)$. It follows that a simple (q,m,k) cover of $V(k,q)$ is the image in $V(k,q)$ of a restriction M of $PG(k-1,q)$ under some representation, where $rk\, M = k$ and $c(M; q) \ge m + 1$. We shall call a restriction of $PG(k-1,q)$ having rank k and critical exponent greater than m a (q,m,k)-*matroid*. A *minimal* (q,m,k)-*matroid* is a (q,m,k)-matroid for which no proper restriction is also a (q,m,k)-matroid.

In [6], Mullin and Stanton proved a number of results for $(2,m,k)$ covers and posed several problems. Most of these were solved in [8] using matroid techniques. The purpose of this paper is to extend these results to matroids representable over an arbitrary finite field.

1. THE (q,m,k)-MATROIDS WITH THE LEAST NUMBER OF ELEMENTS

Let $\eta(q,m,k)$ denote the least number of elements in a (q,m,k)-matroid.

Theorem 1. *(Mullin and Stanton [6, Theorems 4.4, 3.5 and Corollary 1]).*

(i) $\eta(2,m,k) = 2^{m+1} + k - m - 2.$

(ii) *A $(2,2,k)$-matroid having $\eta(2,2,k) = k + 4$ elements is isomorphic to* $PG(2,2) \oplus U_{k-3,k-3}.$

(iii) *A $(2,1,k)$-matroid having $\eta(2,1,k) = k + 1$ elements is isomorphic to* $U_{j,j+1} \oplus U_{k-j,k-j}$ *for some even integer j such that* $2 \le j \le k.$

Mullin and Stanton [6, §5] ask whether (ii) can be extended to $(2,m,k)$-matroids having $\eta(2,m,k)$ elements when $m > 2$. In fact we can extend parts (i) and (ii) of the preceding theorem as follows.

Theorem 2.

(i) $\eta(q,m,k) = \dfrac{q^{m+1}-1}{q-1} + k - m - 1.$

(ii) *A (q,m,k)-matroid having $\eta(q,m,k)$ elements is isomorphic to* $PG(m,q) \oplus$ $\oplus\, U_{k-m-1,k-m-1}$ *for*

(a) $q = 2$ *and* $m \ge 2$; *and*

(b) $q > 2$ *and* $m \ge 1.$

The proof of (ii)(a) was given in [8, Theorem 3] and will not be repeated here. We prove (ii)(b) using a similar method.

We shall use three lemmas.

Lemma 1. $\eta(q,m,k) \leq \dfrac{q^{m+1}-1}{q-1} + k - m - 1.$

Proof. The matroid $PG(m,q) \oplus U_{k-m-1,k-m-1}$ is easily seen to be a (q,m,k)-matroid.

$\square$

Lemma 2. *[8, Lemma 5]. If M is representable over $GF(q)$ and C^* is a cocircuit of M such that $c(M; q) - 1 = c(M \backslash C^*; q) = n$, then $|C^*| \geq q^n$. Hence if M is minimal having critical exponent greater than n, then every cocircuit of M contains at least q^n elements.*

Lemma 3. *(Brylawski [3, Lemma 4.1]). If $r,q > 2$ and $U_{r,n}$ is representable over $GF(q)$, then $c(U_{r,n}; q) = 1$.*

Proof of Theorem 2. We shall assume that $q > 2$ and prove parts (i) and (ii)(b) using induction on m. If $m = 1$ and M is a minimal $(q,1,k)$-matroid, then $c(M; q) = 2$. Let T be the set of coloops of M. Then if $k' = k - |T|$, $M \backslash T$ is a minimal $(q,1,k')$-matroid. Since $M \backslash T$ has no coloops, $M \backslash T$ is minimal having critical exponent 2. Thus by Lemma 2, if C^* is a cocircuit of $M \backslash T$, then $|C^*| \geq q$. We conclude that M has at least $q + k - 1$ elements. Thus, by Lemma 1, $\eta(q,1,k) = q + k - 1$. If M has exactly $\eta(q,1,k)$ elements, then every cocircuit of $M \backslash T$ has exactly q elements and every hyperplane of $M \backslash T$ is a free matroid. Thus $M \backslash T \cong U_{k',k'+q-1}$ and so $M \cong U_{k',k'+q-1} \oplus U_{k-k',k-k'}$. Hence $2 = c(M; q) = c(U_{k',k'+q-1}; q)$ and so, by Lemma 3, $k' \leq 2$. But M is simple and $q > 2$, hence $k' = 2$ and so,

$$M \cong U_{2,q+1} \oplus U_{k-2,k-2} \cong PG(1,q) \oplus U_{k-2,k-2}.$$

Thus both (i) and (ii)(b) hold for $m = 1$.

Assume now that $m > 1$ and that both (i) and (ii)(b) hold for $m - 1$; that is, assume that

(1) $\eta(q,m-1,k) = \dfrac{q^m-1}{q-1} + k - m$; and

(2) every $(q,m-1,k)$-matroid having $\eta(q,m-1,k)$ elements is isomorphic to $PG(m-1,q) \oplus U_{k-m,k-m}$.

Let M be a (q,m,k)-matroid on a set having $\eta(q,m,k)$ elements and let T be the set of coloops of M. If $N = M \backslash T$ and $k' = k - |T|$, then N is a (q,m,k')-matroid having $\eta(q,m,k')$ elements. Thus N is minimal having rank k' and critical exponent $m + 1$. But N has no coloops and hence N is minimal having critical exponent $m + 1$.

Let C^* be a cocircuit of N. Then by Lemma 2, $|C^*| \geq q^m$ and by Lemma 1, N has at

ost $\frac{q^{m+1}-1}{q-1} + k' - m - 1$ elements. Hence $N\backslash C^*$ has at most $\frac{q^m-1}{q-1} + k' - m - 1$ elements. But
\C* has rank $k' - 1$ and critical exponent m, hence $N\backslash C^*$ is a $(q, m-1, k'-1)$-matroid and
o by (1), $N\backslash C^*$ has at least $\frac{q^m-1}{q-1} + k' - m - 1$ elements. Thus $N\backslash C^*$ has exactly $\frac{q^m-1}{q-1} + k' -$
$m - 1$ elements and $|C^*| = q^m$. It follows by (1) and (2) that $N\backslash C^* \cong PG(m-1, q) \oplus$
$U_{k'-m-1, k'-m-1}$. Moreover, since C^* was an arbitrarily chosen cocircuit of N, it
ollows that every cocircuit of N has exactly q^m elements.

If y is a coloop of $N\backslash C^*$, then $y \cup C^*$ contains a cocircuit C_1^* of N containing y.
ince $|C_1^*| = q^m$, $C_1^* = (C^*\backslash y_1) \cup y$ for some y_1 in C^*. Now let z be an element of $C^*\backslash y_1$.
en $z \in C^* \cap C_1^*$ and therefore there is a cocircuit C_z^* of M such that $C_z^* \subseteq (C^* \cup y)\backslash z$. Since
$\mid_z^*\mid = q^m$, it follows that $C_z^* = (C^*\backslash z) \cup y$. We conclude that $C^* \cup y$ is a set of size
$+ 1$ for which every subset of size q^m is a cocircuit. Thus $N .(C^* \cup y) \cong U_{2, q^m+1}$.
t $m > 1$, therefore U_{2, q^m+1} is not representable over $GF(q)$, hence N is not represen-
able over $GF(q)$; a contradiction.

It follows that $N\backslash C^*$ has no coloops; that is, $k' - m - 1 = 0$. Hence $N\backslash C^* \cong PG(m-1, q)$.
t now N is a simple matroid representable over $GF(q)$ having rank $m + 1$ and exactly
$\frac{m+1}{q-1}$ elements. Therefore $N \cong PG(m, q)$ and so $M \cong PG(m, q) \oplus U_{k-m-1, k-m-1}$ and
$(q, m, k) = \frac{q^{m+1}-1}{q-1} + k - m - 1$. □

The next result is an easy consequence of Theorem 2. The following elementary
atroid-theoretic proof of it was communicated privately to the author by Tom Brylawski.

Corollary. *(Bose and Burton [2, Theorem 2]). Let M be a loopless matroid*
presentable over GF(q) and suppose that M has critical exponent greater than m.
en M has at least $\frac{q^{m+1}-1}{q-1}$ elements. Moreover, if M has exactly $\frac{q^{m+1}-1}{q-1}$ elements, then
$\cong PG(m, q)$.

Proof. As M has critical exponent greater than m, $(\lambda-1)(\lambda-q)...(\lambda-q^m)$ divides
$(M; \lambda)$. Now the number of flats of M of rank 1 is equal to the coefficient of
$k M-1$ in $P(M; \lambda)$, hence M has at least $q^m + q^{m-1} + ... + q + 1 = \frac{q^{m+1}-1}{q-1}$ elements.

If M has exactly $\frac{q^{m+1}-1}{q-1}$ elements, then $P(M; \lambda) = (\lambda-1)(\lambda-q)...(\lambda-q^m)$ and hence
has rank $m + 1$. It follows that $M \cong PG(m, q)$, as required. □

An *n-critical* graph [1, p.117] is a graph G having chromatic number n such that
very proper subgraph of G has chromatic number less than n. The next result, a
tural analogue of Lemma 2 for regular matroids, generalizes Dirac's well-known
sult [5, p.45] than an n-critical graph is $(n-1)$-edge-connected. The proof is very
milar to the proof of Lemma 2 as given in [8, Lemma 5]. The *chromatic number* $\chi(M)$
a loopless regular matroid M is the least positive integer j such that $P(M; j) > 0$
ee [9, p.262] or [7]).

(3)   If $1 \le j < \chi(M)$, then $P(M; j) = 0$ (see, for example, [7, Theorem 2.9]).

Theorem 3. *If M is a regular matroid and C^* is a cocircuit of M such that* $\chi(M) = \chi(M \backslash C^*) + 1 = n$, *then* $|C^*| \ge n - 1$. *Hence if M is minimal having chromatic number n, then every cocircuit of M contains at least $n - 1$ elements.*

Proof. By [7, Lemma 2.7], if $\{x_1, x_2, \ldots, x_t\}$ is a cocircuit of M, then

(4) $P(M; \lambda) = (\lambda - t)P(M \backslash x_1, x_2, \ldots, x_t; \lambda)$

$$+ \sum_{j=2}^{t} \sum_{i=1}^{j-1} P(M \backslash x_1, \ldots, x_{i-1}, x_{i+1}, \ldots, x_{j-1}/x_i, x_j; \lambda).$$

If $\chi(M) = \chi(M \backslash x_1, x_2, \ldots, x_t) + 1 = n$, then put $\lambda = n - 1$ in (4). Now, by (3), $P(M; n-1)$ = 0 and each term in the double summation is non-negative. Moreover, $P(M \backslash x_1, x_2, \ldots, x_t; n-1) > 0$. We conclude that $(n-1) - t \le 0$.   □

2.   THE SETS $Q_m(q)$

If $m \in \mathbb{Z}^+$, then let

$$Q_m(q) = \{\text{cork } M: \ M \text{ is a minimal } (q,m,k)\text{-matroid}\}.$$

Mullin and Stanton [6, §5] noted that $Q(2) = \{1\}$ and asked whether $Q_m(2)$ is finite for $m \ge 2$.

Theorem 4. $Q_m(q)$ *contains all but finitely many positive integers provided*

(a) $q = 2$ and $m \ge 2$; or

(b) $q > 2$ and $m \ge 1$.

Part (a) of this theorem was proved in [8, Theorem 1]. It requires only a slight modification of this argument to prove part (b).

Acknowledgements

The author thanks Professor T.A. Dowling for drawing his attention to Bose and Burton's paper, and Professor T.H. Brylawski for several useful discussions. Partial support of this work by CSIRO is also gratefully acknowledged.

References

[1]   J.A. Bondy and U.S.R. Murty, *Graph Theory with Applications* (Macmillan, London; American Elsevier, New York, 1976).

[2]   R.C. Bose and R.C. Burton, A characterization of flat spaces in a finite geometry and the uniqueness of the Hamming and the MacDonald codes. *J. Combinatoria Thoery*, 1 (1966), 96-104.

[3]   Thomas H. Brylawski, An affine representation for transversal geometries. *Studie in Appl. Math.*, 54 (1975), 143-160.

[4]   Henry H. Crapo and Gian-Carlo Rota, *On the Foundations of Combinatorial Theory:*
        Combinatorial Geometries (Preliminary ed., M.I.T. Press, Cambridge, Massachusetts;
        London, 1970).

[5]   G.A. Dirac, The structure of k-chromatic graphs. *Fund. Math.*, 40 (1953), 42-55.

[6]   R.C. Mullin and R.G. Stanton, A covering problem in binary spaces of finite
        dimension. *Graph Theory and Related Topics* (eds. J.A. Bondy and U.S.R. Murty,
        Academic Press, London, New York, San Francisco, to appear).

[7]   James G. Oxley, Colouring, packing and the critical problem. *Quart. J. Math.*
        Oxford (2), 29 (1978), 11-22.

[8]   James G. Oxley, On a covering problem of Mullin and Stanton for binary matroids.
        Aequationes Math. (to appear).

[9]   D.J.A. Welsh, *Matroid Theory* (London Math. Soc. Monographs No. 8, Academic
        Press, London, New York, San Francisco, 1976).

Department of Pure Mathematics, SGS,
and Mathematics, IAS,
Australian National University,
Canberra, A.C.T. 2600,
Australia.

FACTORISATION OF COMPLETE BIPARTITE GRAPHS INTO
TWO ISOMORPHIC SUBGRAPHS

STEPHEN J. QUINN

A construction method is given which generates all factorisations of the complete bipartite graph $K_{m,n}$ into two isomorphic line disjoint subgraphs. Such subgraphs are called self-complementary bipartite subgraphs, by analogy with ordinary self-complementary graphs. It is shown that the factorisation giving rise to a self-complementary bipartite graph is unique up to isomorphism. Based on this fact a method is developed for counting unlabelled self-complementary bipartite graphs.

1. INTRODUCTION

In 1963 Ringel [8] proved that if a graph G was self-complementary then any permutation giving an isomorphism of G to its complement $\overline{G}$ consists entirely of cycles of length 4n (for $n \in \mathbb{N}$) except for at most one fixed point. Later Read [7], using the counting methods of Polya and de Bruijn, enumerated self-complementary graphs with p points.

An *isomorphic factorisation* of a graph G is a partition of its line set E(G) into disjoint isomorphic classes called factors. This concept was recently introduced by Harary, Robinson and Wormald [5]. One obvious necessary condition for an isomorphic factorisation of a graph to exist is that the number of factors divide the number of lines of that graph. This is called the Divisibility Condition.

In particular any pair of self-complementary graphs on p points corresponds to a factorisation of the complete graph K_p into two parts. The object of this paper is the analogous problem of determining all isomorphic factorisations of the complete bipartite graph $K_{m,n}$ into two factors, G and $\overline{G}$, the latter to be termed hereafter the *relative complement* of G. Each factor of $K_{m,n}$ will be called a *self-complementary bipartite graph*. We will adopt the terminology of [5] and say that if such a factorisation exists then G divides $K_{m,n}$ and denote the set of graphs dividing $K_{m,n}$ into two self-complementary bipartite subgraphs by $K_{m,n}/2$. Any graph theoretic terminology and notation not defined in this paper may be found in [2].

In the following section we show that for any isomorphic factorisation of $K_{m,n}$ into two copies of G, the colour classes are unique for G up to isomorphism. This result is used in Section 3 to count the graphs in $K_{n,m}$ for any given n and m.

2. A STRUCTURAL CHARACTERISATION OF SELF-COMPLEMENTARY BIPARTITE GRAPHS

A *colour partition* of a graph is a partition of the point set such that no two adjacent points are in the same colour class in the partition. This is equivalent to a colouring with interchangeable colours. For example, the graph $K_{2,4} \cup 2K_1$ as pictured in Figure 1 has the colour partition $\{a_1 a_2 a_3 a_4\}, \{a_5 a_6 a_7 a_8\}$.

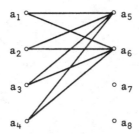

FIGURE 1 - A Self-Complementary Bipartite Graph

It is seen that this partition is induced by an isomorphic factorisation of $K_{4,4}$. The relative complement is isomorphic to the original partition by the permutation $(a_5 a_7)(a_6 a_8)$, which preserves the colour classes as well as the adjacency relation. A bipartite graph equipped with a colour partition having two colours is termed a *bicoloured* graph. The relative complement of a bicoloured graph G is induced on the point set of G by including every line between points in opposing colour classes which are not already adjacent in G.

In general an isomorphic factorisation of a graph with a colour partition into copies of a graph induces colour partitions on G which may not be isomorphic among these copies, or may not be isomorphic to colour partitions induced by some other isomorphic factorisation. In the case of the former we display, in Figure 2, an isomorphic factorisation of $K_{3,2}$ into three parts where the factors are not all isomorphic as bicoloured graphs.

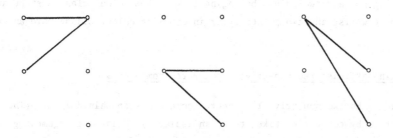

FIGURE 2 - A Factorisation of $K_{3,2}$ into Three Parts

The latter is illustrated in Figure 3 with two distinct factorisations of the complete tripartite graph $K_{1,2,2}$ into two copies of the path P_4. In the first factorisation the two copies of P_4 with the inherited colour classes are isomorphic to each other, but not to those obtained in the second factorisation, and vice versa.

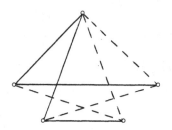 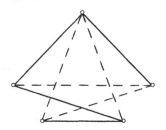

FIGURE 3 - Two Factorisations of $K_{1,2,2}$ into Two Copies of P_4

As we now show, this anomaly never occurs in self-complementary bipartite graphs.

Theorem. *Let G and G' respectively be factors in two isomorphic factorisa-tions of complete bipartite graphs into two parts, and let G and G' inherit colour partitions from their respective factorisations. Then G and G' are isomorphic as bicoloured graphs if the underlying graphs (without the colour partitions) are iso-morphic.*

Proof. Consider any isomorphic factorisation of a complete bipartite graph into two parts which we denote by G and $\overline{G}$. The proof is divided into three cases and in each it is shown that the colour classes for G are unique up to isomorphism.

Case 1. G is connected

Let ϕ be an isomorphism which maps G to its relative complement. It follows immediately that the two colour classes of G are uniquely determined since ϕ is distance preserving and any two points belonging to the same colour class must be an even distance apart whilst any two points lying in opposing colour classes are an odd distance apart.

Case 2. G contains at least two non-trivial connected components

Let G_1 and G_2 be two non-trivial connected components contained in G. (The trivial component is by convention taken to be an isolate.) Using the argument given in Case 1, the colour classes of the points comprising G_1 and G_2 are uniquely deter-mined up to isomorphism and so, without loss of generality, we represent the points in

G_1 by $\{a_1,a_2,\ldots,a_m\}$, $\{b_1,b_2,\ldots,b_n\}$ and the points in G_2 by $\{c_1,c_2,\ldots,c_s\}$, $\{d_1,d_2,\ldots,d_t\}$ and denote these point sets by A, B, C and D respectively. Without loss of generality we shall assume that the points belonging to A and C lie in the same colour class of G, and therefore that the points in B and D are contained in the opposing colour class of G. Now G_1 and G_2 are disjoint in G and therefore any two points in A and D or B and C are adjacent in $\overline{G}$. Hence there are two complete bipartite subgraphs $K_{s,n}$ and $K_{m,t}$ embedded in $\overline{G}$. Moreover, if any two points belonging to A and B or C and D are not adjacent in G, or G contains any other components, then $\overline{G}$ is connected which contradicts our initial assumptions. Therefore $G \cong K_{m,n} \cup K_{s,t}$ and $\overline{G} \cong K_{m,t} \cup K_{s,n}$. Since G and $\overline{G}$ contain the same number of lines then either s = m or t = n.

It follows directly that the two colour classes of G, namely $\{a_1,a_2,\ldots,a_m,c_1,c_2,\ldots,c_s\}$ and $\{b_1,b_2,\ldots,b_n,d_1,d_2,\ldots,d_t\}$ are unique up to isomorphism.

Case 3. G contains exactly one non-trivial connected component together with some isolates

Let G_1 be the non-trivial connected component of G. From Case 1 it follows that the points of G_1 have a unique colour partition into two classes, A = $\{a_1,a_2,\ldots,a_n\}$ and B = $\{b_1,b_2,\ldots,b_m\}$ say. Let $\{c_1,c_2,\ldots,c_q\}$ represent the set of isolates of G, and denote this set by C.

In the course of the proof we will need the following notion first defined in [4]. A bicoloured graph H is *symmetric* if there exists an automorphism which reverses the colour classes of H. If H is not symmetric, then we say that H is *non-symmetric*. One obvious necessary condition for a graph to be symmetric is that the numbers of points in each colour class are equal.

We shall show that every isolate can belong to one and only one colour class of G, unless G is symmetric. Clearly, G cannot contain isolates in both colour classes since the relative complement would be connected. When G_1 is symmetric it follows by definition that the colour classes of G are unique up to isomorphism. We consider first a situation in which we are able to show that G_1 is symmetric.

Case 3A. There is an isomorphism ϕ from G to its relative complement which does not preserve the colour classes of G.

Without loss of generality, assume that G has colour classes A ∪ C and B. We consider first the case when C contains just one isolate, which we denote by c_1. Then ϕ maps c_1 to a point in A, the points in A to the points in B, and the points in B to c_1 and the remaining points in A. This is seen by recalling that two points in

opposing colour classes of G_1 are an odd distance apart and hence are mapped to points
which lie an odd distance apart in $\overline{G}$. Moreover, if ϕ maps c_1 to a point in B, then at
least one point in B and one point in A are mapped to points in A ∪ C, which contradicts the preceeding statement. Hence ϕ maps c_1 to a point in A. Moreover, since ϕ
does not preserve the colour classes of G, then it follows that the points in A are
mapped to the points in B, and the points in B are mapped to c_1 and the remaining
points in A. Without loss of generality suppose that ϕ maps c_1 to a_1, a_1 to b_1 and,
in general, the points a_k to b_k, b_k to a_{k+1} for $k < n$. Using induction we shall show
for all $i, j \leq m$ that $a_i \in A$, $b_j \in B$, a_i has degree $n + 1 - i$, b_j has degree j, and that G_1
is isomorphic to the graph shown in Figure 4. In this graph a_i is adjacent to b_j if
and only if $i \leq j$.

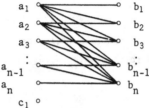

FIGURE 4   A self-complementary bipartite graph

Consider the isolate c_1 in C. As shown earlier c_1 is mapped to $a_1 \in A$. Since
in G the degree of c_1 is zero and ϕ preserves the adjacency relation of the lines of G,
it follows that the degree of a_1 in G is zero and hence in $\overline{G}$ is n. Therefore a_1 is
adjacent to every point in B. Now a_1 under ϕ is mapped to b_1, which is contained in B
since ϕ maps the points in A to the points in B. Further b_1 is adjacent to n of the
points in A ∪ C in $\overline{G}$ and hence to one point, namely a_1, in G. Similarly ϕ maps b_1 to
a_2, which therefore has degree 1 in $\overline{G}$, and $n - 1$ in G. Hence $a_2 \in A$, unless $n = 1$, in
which case $a_2 = c_1$ and G_1 is isomorphic to the symmetric component K_2.

We proceed by induction on k for $1 \leq k < n$ and assume for all $i, j \leq k$ in G
that $a_i \in A$, $b_j \in B$, a_i has degree $n + 1 - i$, b_j has degree j, and that a_i and b_j are
adjacent if and only if $i \leq j$.

Consider the point b_k in G. Now ϕ maps b_k to a_{k+1} and thus it follows by our
induction assumption that in $\overline{G}$, a_{k+1} has degree k, and hence has degree $n - k > 0$ in G.
Therefore, $a_{k+1} \in A$. Similarly, ϕ maps a_{k+1} to b_{k+1}, and therefore b_{k+1} has degree
$n - k$ in $\overline{G}$ and $k + 1$ in G. Moreover, since $a_{k+1} \in A$ and ϕ maps the points in A to those
in B it follows that $b_{k+1} \in B$.

It still remains to show for $i, j \leq k + 1$ that a_i and b_j are adjacent in G if and
only if $i \leq j$. As before we consider b_k. By our induction assumption b_k is adjacent
to the points $a_i \in A$ in G for $i \leq k$. Hence in $\overline{G}$, a_{k+1} is adjacent to $b_i \in B$ for $i \leq k$;
therefore in G, a_{k+1} is not adjacent to $b_i \in B$ for $i \leq k$. We now focus our attention on
b_{k+1}. By similar reasoning to the above, b_{k+1} is not adjacent to $a_i \in A$ in $\overline{G}$ for
$2 \leq i \leq k + 1$. Therefore in G, b_{k+1} is adjacent to a_i for $2 \leq i \leq k + 1$. Furthermore, since
a_1 is adjacent to every point in B in G, it follows that b_{k+1} and a_i are adjacent for

for all $i \leqslant k+1$. Hence, it follows for $i,j \leqslant k+1$ that a_i and b_j are adjacent in G if and only if $i \leqslant j$.

Lastly we consider the point $b_n \in B$ in G. By our induction b_n has degree n. Further ϕ maps b_n to a point in $A \cup C$ in $\overline{G}$ with degree n, and therefore in G this point has degree zero. Hence ϕ maps b_n to c_1 which completes the induction.

Suppose now that C contains q isolates where $q > 1$. Then the same argument suffices to show that the points in A and B can each be divided into n/q parts, say $A_1, A_2, \ldots, A_{n/q}$ and $B_1, B_2, \ldots, B_{n/q}$, and hence that G_1 is symmetric. In this instance, two points $a \in A_i$ and $b \in B_j$ are adjacent in G if and only if $i \leqslant j$.

Thus G_1 is symmetric as seen by considering the permutation ψ which maps the points in A_j in some order to the points in $B_{n/q+1-j}$ and vice versa for $j = 1, 2, \ldots, n/q$.

<u>Case 3B</u>. Every isomorphism ϕ from G to $\overline{G}$ preserves the colour classes of G.

Again we suppose without loss of generality that the colour classes of G consist of $A \cup C$ and B, where C contains just one isolate, which we denote by c_1. In this case ϕ maps c_1 to a point a_1 say, in A, the points in A to c_1 and the remaining points in A, and the points in B to themselves. We shall show by induction that if G is non-symmetric then the colour partition of G is unique up to isomorphism.

As before, we consider the point c_1 in C. Since ϕ maps c_1 to a_1, it follows that there exists a complete bipartite subgraph $K_{1,n}$ embedded in G with line set $\{a_1, b_j\}$ for $j = 1, 2, \ldots, n$. If $n = 1$ then $G_1 \cong K_2$ which is symmetric and falls under Case 3A. Hence we further assume that $n \neq 1$. Suppose, on the other hand, that in G c_1 is taken to belong to the same colour class as the points in B. Let G' be the bipartite graph obtained from $G_1 \cup C$ by designating the colour classes A and $B \cup C$, and suppose that G' is relatively self-complementary. Then there exists an isomorphism ϕ', say, from G to its relative complement G'. If ϕ' does not preserve the colour classes of G' then from the first part of Case 3, G_1 is symmetric, which contradicts the assumption that every map ϕ from G to $\overline{G}$ preserves the colour classes of G. Hence every such isomorphism ϕ' maps c_1 to a point in B, without loss of generality the point b_1, the points of B to c_1 and the remaining points in B, and the points in A to themselves.

By similar reasoning G_1 contains a second complete bipartite subgraph $K_{n,1}$ with line set $\{a_i, b_1\}$ for $j = 1, 2, \ldots, n$. Hence, in G', ϕ' maps the point a_1 to a point in A of degree n. Moreover, since $n \neq 1$, it follows that ϕ' does not fix a_1, and therefore we assume without loss of generality that ϕ maps a_1 to a_2. Similarly in G we shall assume that ϕ maps the point b_1 to the point b_2.

Therefore $n \geqslant 2$, and hence for $n < 2$ the colour classes of G are unique up to isomorphism, that is, if G and its relative complement $\overline{G}$ are isomorphic then G' and $\overline{G}'$ are not isomorphic. We proceed by induction and assume that the colour classes are unique for any self-complementary bipartite graph in $K_{p,p+1/2}$ containing one isolate and one non-trivial connected component.

Consider the $(n-1)(n-2)$ lines between the points a_i and b_j for $i = 2,3,\ldots,n$ and $j = 3,4,\ldots,n$, still to be apportioned between the factors G and $\overline{G}$. Let H be the induced subgraph of G, having colour classes $\{a_i \mid i = 2,3,\ldots,n\}$ and $\{b_j \mid j = 3,4,\ldots,n\}$. It is clear that the line set E of H contains exactly $\frac{(n-1)(n-2)}{2}$ lines. Moreover, it follows by the symmetry of the construction so far that H is a relatively self-complementary bipartite graph in $K_{n-1,n-2/2}$. Similarly, we let H' be the induced subgraph of G' with colour classes $\{a_i \mid i = 3,4,\ldots,n\}$ and $\{b_j \mid j = 2,3,\ldots,n\}$, and line set E. Then it follows that G' and $\overline{G}'$ are relatively self-complementary if and only if $H' \in K_{n-2,n-1/2}$.

Consider the subgraphs H and H' embedded in G and G' respectively. Since H and H' have the same line set E but different point sets, then it follows that H contains at least one isolate, namely a_2. Similarly, in H' the point b_2 is isolated. Moreover, if H contains any other isolates then H' is connected. Hence H contains exactly one isolate a_2 and one connected component E. Then by our induction assumption it follows that the colour classes of H are unique up to isomorphism. This, however, contradicts the fact that $H' \in K_{n-2,n-1/2}$. Hence it follows that G' and $\overline{G}'$ are not isomorphic, and therefore that the colour classes of G are uniquely determined up to isomorphism.

In similar spirit to Case 3A, if C contains q isolates where $q > 1$, then the same argument suffices to prove that the colour classes of G are uniquely determined up to isomorphism.

Corollary 1.   *Let $\{G_1,\overline{G}_1\}$ and $\{G_2,\overline{G}_2\}$ be factorisations of complete bipartite graphs such that the four factors are isomorphic as ordinary (uncoloured) graphs. Then there is an isomorphism from the first factorisation to the second which maps G_1 to G_2.*

Proof.   By the previous theorem there is an isomorphism ϕ mapping G_1 to G_2 which preserves the colour classes of G_1. Since any factorisation is completely determined by the colour classes of one of its factors it follows directly that ϕ is the required isomorphism which maps the first factorisation to the second.

Corollary 2.   *Let G_1 and $\overline{G}_1$ be factors in an isomorphic factorisation of $K_{m,n}$ into two parts. Then there exists an isomorphism ϕ which maps G_1 to its relative complement which preserves the colour partition of G_1. Moreover, if ϕ preserves the colour classes of G_1 then every cycle involving points from one of the colour classes has even length, and if ϕ reverses the colour classes of G_1 then every cycle in ϕ has doubly even length.*

Proof.   Consider two copies of the factorisation $\{G_1, \overline{G}_1\}$. Since the underlying graphs of G_1 and $\overline{G}_1$ are isomorphic, it follows by the previous theorem that G_1 and $\overline{G}_1$ are isomorphic as bicoloured graphs, that is, by an isomorphism which preserves the colour partition.

We now turn our attention towards analysing the above isomorphisms. When the two colour classes of G_1 are preserved by ϕ the lines of $\overline{G}_1$ are then necessarily mapped to those of G_1 and vice versa. Any cycle of length α consisting of points belonging to one of the colour classes of G_1 together with any cycle of length β containing points in the opposite colour class of G_1 will induce exactly (α, β) cycles each of length $[\alpha, \beta]$, where (α, β) and $[\alpha, \beta]$ are respectively the greatest common divisor and least common multiple of α and β. Each of the (α, β) line cycles consist of lines lying alternately in G and $\overline{G}$ and so $[\alpha, \beta]$ must be even. This implies that at least one of α or β is even. As this is the case for any pair of cycles with points in opposing colour classes, then it follows that every cycle containing points from one of the colour classes must have even length.

Similarly when ϕ reverses the two colour classes of G_1 then every cycle length $4m$ (for $m \in \mathbb{N}$) since the diagonal line cycle induced by any cycle of length $4m + 2$ has odd length while containing lines lying alternately in G_1 and $\overline{G}_1$, which is a contradiction. Figure 5 shows two factors of $K_{6,6}$ illustrating the different types of permutations. Two permutations which induce the relative complements of the graphs pictured in Figures 5(a) and 5(b) are given by $(b_1b_6)(b_2b_5)(b_3b_4)$ and $(a_1b_1a_3b_3)(a_2b_2a_4b_4)(a_5b_5a_6b_6)$ respectively.

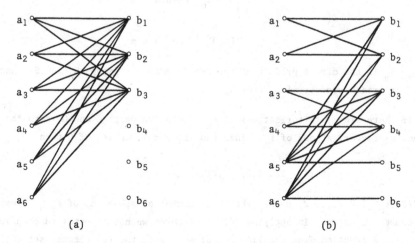

(a)   (b)

FIGURE 5 - Two self-complementary bipartite graphs

3. THE NUMBER OF GRAPHS, G, IN $K_{m,n}$

The cycle index of a group G of degree n and order $|G|$ is defined to be the polynomial

$$Z(G;c_1,c_2,...,c_n) = \frac{1}{|G|} \sum_{(j)} n_{j_1,j_2,...,j_n} c_1^{j_1} c_2^{j_2},...,c_n^{j_n}$$

where $c_1,c_2,...,c_n$ are variables; $n_{j_1,j_2,...,j_n}$ is the number of permutations of G of type $(j_1,j_2,...,j_n)$ where j_i denotes the number of cycles of length i; and the summation over (j) is such that $j_1 + 2j_2 + ... + nj_n = n$. When it is unnecessary to display the variables $c_1,c_2,...,c_n$ we shall write $Z(G)$ instead of $Z(G;c_1,c_2,...,c_n)$.

In 1963 Read [7] counted all self-complementary graphs on n points. He proved that the total number of all such graphs is given by $Z(S_p^{(2)};0,2,0,2,...,0,2)$ where $S_p^{(2)}$ denotes the pair group of S_p. That is, $S_p^{(2)}$ is the representation of S_p as permutations on the $\binom{p}{2}$ unordered pairs from a p-element set.

To count the number of self-complementary bipartite graphs we use a well known result of Pólya: that the number of non-isomorphic subgraphs of $K_{m,n}$ with k edges is the co-efficient of x^k in

$$Z\left(\Gamma_1(K_{m,n}); 1+x,1+x^2,...,1+x^{nm}\right)$$

where $\Gamma_1(K_{m,n})$ denotes the line group of $K_{n,m}$. A detailed exposition of Pólya's Haupsatz can be found in [3,Chapter 2]. Harary [1] has shown that

$$Z\left(\Gamma_1(K_{m,n})\right) = \begin{cases} S_n \times S_m & \text{when } n \neq m \\ \\ [S_n]^{S_2} & \text{when } n = m \end{cases}$$

where $S_n \times S_m$ is the direct product of the two permutation groups S_n and S_m, and $[S_n]^{S_2}$ denotes the wreath product of S_n over S_2.

If there are g_1 factorisations of $K_{m,n}$ into two parts G and $\overline{G}$, such that $G \not\cong \overline{G}$ and g_2 factorisations of $K_{m,n}$ into two isomorphic copies of G then

$$Z\left(\Gamma_1(K_{m,n});2,2,2,...,2\right) = 2g_1 + g_2 \qquad (1)$$

since $Z\left(\Gamma_1(K_{m,n});2,2,,...,2\right)$ counts the total number of subgraphs of $K_{m,n}$ irrespective of the number of edges. In applying Pólya's Theorem we have identified each subgraph of $K_{m,n}$ with a function from the line set of $K_{n,m}$ into the two element set $\{0,1\}$. This function takes the value 1 on a line contained in a subgraph and 0 otherwise. Two functions are counted as equivalent if their corresponding subgraphs are isomorphic by one of the automorphisms of $K_{m,n}$. This notion of equivalence may be generalised by identifying a graph with its relative complement and under this generalisation the number of inequivalent subgraphs of $K_{m,n}$ is $g_1 + g_2$. These graphs may be counted using de Bruijn's method (see [3,Chapter 6]). In this manner we obtain the left hand side of equation (2)

$$\tfrac{1}{2}Z\big(\Gamma_1(K_{m,n});0,2,0,2,\ldots,0,2\big) + \tfrac{1}{2}Z\big(\Gamma_1(K_{m,n});2,2,\ldots,2\big) = g_1 + g_2. \tag{2}$$

From equations (1) and (2) the total number of self-complementary bipartite graphs is given by

$$Z\big(\Gamma_1(K_{m,n});0,2,0,2,\ldots,0,2\big) = g_1. \tag{3}$$

To evaluate (3) we consider two cases.

Case 1. The colour classes of $K_{m,n}$ are unequal in size

Harary [1] has shown that

$$Z(S_n \times S_m; c_1, c_2, \ldots, c_{nm}) = Z(S_n; a_1, a_2, \ldots, a_m) \times Z(S_m; b_1, b_2, \ldots, b_n). \tag{4}$$

where $a_\alpha^{i_\alpha} \times b_\beta^{j_\beta} = c_{[\alpha,\beta]}^{i_\alpha j_\beta (\alpha,\beta)}$ and $[\alpha,\beta]$, (α,β) are, as before, the least common multiple and greatest common divisor of α and β. The operator 'x' is bilinear on $Q[a_1,a_2,\ldots,a_m]$, $Q[b_1,b_2,\ldots,b_n]$ and distributes over products on both sides. For example, $a_5^2 a_6 \times b_4^3 = c_{20}^6 c_2^6$.

We illustrate by determining the number of graphs in $K_{2,4}/2$.

$$Z(S_2) = \tfrac{1}{2}(a_1^2 + a_2)$$

$$Z(S_4) = \frac{1}{24}(b_1^4 + 3b_2^2 + 6b_1^2 b_2 + 6b_4 + 8b_1 b_3)$$

$$Z(S_2 \times S_4) = \frac{1}{48}(c_1^8 + 6c_1^4 c_2^2 + 8c_1^2 c_3^2 + 8c_2 c_6 + 12c_4^2 + 13c_2^4)$$

and therefore $Z(S_2 \times S_4; 0,2,0,2,\ldots,0,2) = 6$. The six self-complementary bipartite graphs in $K_{2,4}/2$ are pictured in Figure 6.

FIGURE 6 - The set of graphs in $K_{2,4}/2$

Case 2. The colour classes of $K_{m,n}$ are equal in size
_____

It was shown in [1] that when n = m

$$Z\big(\Gamma_1(K_{n,n})\big) = \tfrac{1}{2}Z(S_n \times S_n) + \tfrac{1}{2}Z'_n \tag{5}$$

where $Z(S_n \times S_n)$ is defined as in Case 1, and

$$Z'_n = \frac{1}{n!} \sum_{(j)} \frac{n!}{\prod k^{j_k} \cdot j_k!} \cdot \prod_{k \text{ even}} c_k^{j_k/2 + k\binom{j_k}{2}} \cdot \prod_{k \text{ odd}} (c_k c_{2k}^{(k-1)/2})^{j_k} c_{2k}^{k\binom{j_k}{2}} \cdot$$

$$\cdot \prod_{r<s} c_{[r,s]}^{j_r j_s (r,s)} \quad .$$

By omitting all terms containing c_i with i odd and then replacing all the remaining c_i's by 2 we obtain

$$Z'_n(0,2,0,2,\ldots,0,2) = \sum_{(j)} \frac{1}{\prod\limits_{k \text{ even}} k^{j_k} \cdot j_k!} \, 2^p \tag{6}$$

where

$$p = \sum_{k \text{ even}} k j_k^2/2 + \sum_{1 \leq r < s \leq n} j_r j_s (r,s) \tag{7}$$

Writing r = 2α, s = 2β, $j_{2k} = v_k$ and N = 2n reduces (6) to

$$p = \sum_{k=1}^{N} k v_k^2 + 2 \sum_{1 \leq \alpha < \beta \leq N} v_\alpha v_\beta (\alpha, \beta) \quad . \tag{8}$$

From equations (4), (6) and (8) the number of self-complementary bipartite graphs for small n and m are readily calculated and are displayed in Table 1.

n \ m	1	2	3	4	5	6
1	0	1	0	1	0	1
2		2	3	6	5	19
3			0	7	0	14
4				16	37	114
5					0	196
6						649

TABLE 1 - Numbers of self-complementary bipartite graphs for small m and n

The question arises as to exactly what type of graphs $Z(S_n \times S_n; 0,2,\ldots,0,2)$ and $Z_n'(0,2,\ldots,0,2)$ enumerate. In dealing with this question we shall show that a bipartite graph G is symmetric and relatively self-complementary if and only if there exist isomorphisms ϕ_1 and ϕ_2 say, which map G to its relative complement by respectively reversing or preserving the colour classes of G.

Consider the symmetric graph G and let ψ be an automorphism which reverses the colour classes of G. Further, let θ be an isomorphism which maps G to its relative complement. If θ preserves the colour classes of G then $\theta\psi$ is the required automorphism which reverses the colour classes of G. Similarly, if θ reverses the colour classes of G then $\theta\psi$ preserves the colour classes of G. Conversely, if ϕ_1 and ϕ_2 are isomorphisms which map G to $\overline{G}$ by respectively reversing and preserving the colour classes of G, then $\phi_1^{-1}\phi_2$ is the required automorphism of G which reverses the colour classes of G.

From the above result it follows that if G is not symmetric then every isomorphism from G to $\overline{G}$ either preserves the colour classes of G, or reverses the colour classes of G. We shall call G *strict* if the only isomorphisms which map G to $\overline{G}$ preserve the colour classes of G, and *reversible* if the colour classes of G are reversed. Any relatively self-complementary bipartite graph can then be described as either symmetric, strict or reversible. In Figures 7, 5(a) and 5(b), we exhibit three graphs that are respectively symmetric, strict and reversible.

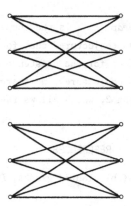

FIGURE 7 - A symmetric graph in $K_{6,6}/2$

Suppose that $K_{m,n}$ contains s self-complementary bipartite subgraphs of which s_1 are symmetric, s_2 are strict and s_3 are reversible. Then

$$Z(S_n \times S_n; 0,2,0,2,\ldots,0,2) = s_1 + 2s_2 , \qquad (9)$$

and

$$Z_n'(0,2,0,2,\ldots,0,2) = s_1 + 2s_3. \qquad (10)$$

Equation (9) follows from the fact that $Z(S_n \times S_n)$ counts the number of graphs in which the relative complement is induced by a permutation that preserves the colour classes of G. In like manner to equation (1), each strict graph is counted as being inequivalent to the graph obtained by reversing the colour classes, whereas each symmetric graph is counted as being equivalent to the graph obtained when the colour classes are reversed. Clearly there is no contribution from a graph that is reversible since every permutation which maps G to its relative complement does so by reversing the colour classes of G.

Equation (10) follows immediately from equations (3) and (5), and the fact that the total number of self-complementary bipartite graphs is

$$g_1 = s_1 + s_2 + s_3 .$$

4. OPEN PROBLEMS

In this section we pose two problems related to the results given in Sections 2 and 3.

A. It was shown in Section 3 that every self-complementary bipartite graph G, say, containing colour classes of equal size n could be described as being symmetric, strict or reversible. We ask for some method which determines the numbers of graphs of each of the three types for given n.

B. The *complete equipartite graph* $K_{n;q}$ is the complement of the union of q disjoint copies of K_n. In [6] it was conjectured that for given n,q and t the Divisibility Condition is sufficient for the existence of a graph in $K_{n;q/t}$. The author has recently verified this conjecture. We therefore ask for an analogous theorem to the one given in Section 2, which allows the enumeration of self-complementary equipartite graphs.

REFERENCES

[1]   F. Harary, On the number of bicoloured graphs, *Pacific J. Math.*, 8 (1958), 743-755.

[2]   F. Harary, *Graph Theory* (Addison-Wesley, Mass., 1969).

[3]   F. Harary and E.M. Palmer, *Graphical Enumeration* (Academic Press, New York, 1973).

[4]   F. Harary and G. Prins, Enumeration of bicolourable graphs, *Canad. J. Math.* 15 (1963), 237-248.

[5]   F. Harary, R.W. Robinson and N.C. Wormald, Isomorphic factorisations I: complete graphs, *Trans. Amer. Math. Soc.*, 242 (1978), 243-260.

[6]   F. Harary, R.W. Robinson and N.C. Wormald, Isomorphic factorisations III:
        complete multipartite graphs, *Proc. Int. Conf. on Combinatorial
        Theory* (Springer-Verlag, Berlin, to appear).

[7]   R.C. Read, On the number of self-complementary graphs and digraphs, *J. London
        Math. Soc.*, 38 (1963), 99-104.

[8]   G. Ringel, Selbstkomplementäre graphen, *Arch. Math.*, 14 (1963), 354-358.

DECOMPOSITION OF INTEGRAL PSEUDOMETRICS

D. F. ROBINSON

Integral pseudometrics arise in a certain type of numerical taxonomy as a dissimilarity coefficient. To each partition on a finite set corresponds a pseudometric on the set in which the distance between points in the same subset is zero, and between points in different subsets is one. Each dissimilarity coefficient is a sum of partition pseudometrics.

We may naturally then ask whether given a dissimilarity coefficient we can recover the partitions, and whether every integral pseudometric on a finite set can be a dissimilarity coefficient. We answer both these questions in the negative.

We give a necessary condition for a decomposition into partition pseudometrics where every partition has two subsets, and an example illustrating the method of attempting a decomposition.

Numerical taxonomy is concerned with formal methods of classification. A finite set A of objects (taxa) is to be classified. We explore the difference between members of A by asking a number of questions, each about a certain character. For instance we might be describing a group of plants, so characters might be leaf shape, leaf texture, petal colour.

Each pair of taxa scores 1 for each character in which they differ and 0 for each character in which they agree. Adding over all characters gives a <u>dissimilarity coefficient</u> for the elements of set A. Further knowledge of numerical taxonomy is not necessary for understanding this paper, but for further discussion see [1] or [3].

Any dissimilarity coefficient is an <u>integral pseudometric</u>. That is, it is a function ρ: $A \times A \to \mathbf{Z}$ with the properties:

(i) $\rho(x,y) \geqslant 0$ for all $x,y \in A$

(ii) $\rho(x,x) = 0$ for all $x \in A$

(iii) $\rho(x,y) = \rho(y,x)$ for all $x,y \in A$

(iv) $\rho(x,y) + \rho(y,z) \geqslant \rho(x,z)$ for all $x,y,z \in A$

If also $\rho(x,y) = 0$ only when $x = y$, we say that ρ is a <u>metric</u>. See, for example, [2].

The following results about pseudometrics are easily established:

THEOREM 1: If ρ is a pseudometric on a set A and $\rho(x,y) = 0$, then $\rho(x,z) = \rho(y,z)$ for all $z \in A$.

THEOREM 2: If ρ and σ are integral pseudometrics on a set A and we define

$(\rho + \sigma)(x,y) = \rho(x,y) + \sigma(x,y)$, then $(\rho + \sigma)$ is also an integral pseudo metric on A.

We introduce also the notation $n\rho$ to represent the result of adding n terms, all equal to ρ, or, alternatively,

$$(n\rho)(x,y) = n.(\rho(x,y)).$$

The zero pseudometric, ω, is defined by

$$\omega(x,y) = 0 \text{ for all } x,y \in A$$

Addition of integral pseudometrics is easily seen to be commutative and associative, and ω acts as an algebraic zero.

Returning to our introductory example, each character defines a partition P on A according to the various states observed for the character. Thus we may define for any partition P on a finite set A a <u>partition pseudometric</u> π_P such that

$\pi_P(x,y) = 0$ if x and y are in the same subset in P.

$\pi_P(x,y) = 1$ if x and y are in different subsets in P.

A special case is the trivial partition {A}: its pseudometric is ω. We will henceforth consider only the pseudometrics belonging to nontrivial partitions.

Then the dissimilarity coefficient may be considered as a sum of suitable partition pseudometrics;

$$\rho = n_1\pi_1 + n_2\pi_2 + \ldots + n_m\pi_m.$$

Suppose we are given the resulting dissimilarity coefficient. Can we always reconstruct the partition pseudometrics of which it is the sum? In other words, is the decomposition of a dissimilarity coefficient into partition pseudometrics unique? Further, if we start with an integral pseudometric, can the decomposition be done at all?

A further specialisation of this problem arises from the fact that in many cases the partitions used have only two subsets, corresponding to the presence and absence of some character. A partition P of a set A into two non-empty subsets will be called a <u>bipartite partition</u>, and the corresponding pseudometric a <u>bipartite pseudo-metric.</u> We may thus also ask whether an integral pseudometric is a sum of bipartite pseudometrics. In this paper we report partial success in answering these questions.

For later use it is now convenient to define a <u>restriction</u> of a pseudometric. If B is a non-empty subset of A and ρ is a pseudometric on A then we define ρ_B by

$$\rho_B(x,y) = \rho(x,y) \quad \text{for all } x,y \in B$$

It can be shown immediately that if ρ is a partition pseudometric corresponding to the partition

$$P = \{A_1, A_2, \ldots, A_m\} \text{ of A}$$

then $\qquad P_B = \{A_1 \cap B, A_2 \cap B, \ldots, A_k \cap B\}$ of B

in which we consider the empty subsets to be ignored. It follows that if P is a bipartite partition then P_B is also bipartite unless B is contained in one of the subsets of P, when P_B is the trivial partition $\{B\}$.

An interesting theorem at this point is:

THEOREM 3: Let ρ be an integral pseudometric such that $\rho(x,y) \in \{0,1\}$ for all $x,y \in A$. Then ρ is a partition pseudometric.

PROOF: We define first a relation R on A by x R y if $\rho(x,y) = 0$.

By (ii) of the definition R is reflexive, by (iii) it is symmetric, and by Theorem 1 it is transitive. Thus R is an equivalence relation and the corresponding partition has ρ as its pseudometric.

One class of bipartite partitions is useful for the construction of examples. For each $r \in A$ we define

$$\beta_r(x,y) = 0 \text{ if } x = y$$
$$= 1 \text{ if } x = r, \ y \neq r;$$
$$= 1 \text{ if } y = r, \ x \neq r;$$
$$= 0 \text{ otherwise.}$$

Also we may define the special pseudometric α

$$\alpha(x,y) = 1 \text{ if } x \neq y$$
$$= 0 \text{ if } x = y;$$

α corresponds to the partition of A into singletons.

THEOREM 4: $\sum_{r \in A} \beta_r = 2\alpha$

The PROOF is by checking for each $x,y \in A$. Only two terms of $\sum_{r \in A} \beta_r(x,y)$ are non-zero for given x and y.

This counterexample shows that there is sometimes more than one collection of partition pseudometrics with a given sum.

Next comes the existence. If A has two members, then there is only one partition pseudometric, which may be described as α, and every other pseudometric is of the form $n\alpha$ for a suitable nonnegative integer n.

When A has three members, let $A = \{1,2,3\}$. Write $d(2,3) = a$, $d(1,3) = b$, $d(1,2) = c$. The available partition pseudometrics are β_1, β_2, β_3 and α. Considering the general sum

$$\rho = r\beta + s\beta_2 + t\beta_3 + u\alpha,$$

we obtain a set of equations which has the solution

$$r = \tfrac{1}{2}(b + c - a) - \tfrac{1}{2}u$$
$$s = \tfrac{1}{2}(c + a - b) - \tfrac{1}{2}u$$
$$t = \tfrac{1}{2}(a + b - c) - \tfrac{1}{2}u$$

Now all the expressions like $(b + c - a)$ are non-negative and all odd or all even, hence there is a solution, obtained by putting u = 0 if a + b + c is even and u = 1 if a + b + c is odd.

Thus for A having 2 or 3 members every pseudometric is a dissimilarity coefficient. However, trying sets of four elements so on yields a counter-example. If there are four elements there are fourteen partition pseudometrics and it is soon seen that no sum can yield the pseudometric:

$$\rho(1,1) = \rho(2,2) = \rho(3,3) = \rho(4,4) = 0$$

$$\rho(1,2) = 2$$

$$\rho(x,y) = 1 \text{ otherwise.}$$

After some experimentation we find that it is the small size of the numbers in this counterexample that cause the failure. There is no room for manoeuvre.

The situation in trying to represent ρ as a sum of bipartite pseudometrics is in some ways easier than the wider problem, for it is easy to establish a necessary condition.

THEOREM 5: If ρ is a bipartite pseudometric on a set A, and x,y,z are three distinct members of A, then $\rho(x,y) + \rho(y,z) + \rho(z,x)$ is even.

PROOF: Either all of x,y,z are in the same subset of the partition, or one is in one subset and two are in the other. In the former case all three distances are zero. In the latter, suppose x is in one subset and y and z in the other. Then

$$\rho(x,y) = \rho(z,x) = 1 \text{ and } \rho(y,z) = 0.$$

Then $\rho(x,y) + \rho(y,z) + \rho(z,x) = 2$.

COROLLARY 6: If an integral pseudometric ρ on a set A is a sum of bipartite pseudometrics then

$$\rho(x,y) + \rho(y,z) + \rho(z,x)$$

is even for all sets $\{x,y,z\}$ of points in A.

COROLLARY 7: If an integral pseudometric ρ on a set A is a sum of bipartite pseudometrics then either two of, or none of ,

$$\rho(x,y), \ \rho(y,z), \ \rho(z,x)$$

are odd.

We now define the <u>odd-distance graph</u> G_ρ of a pseudometric on A. This graph has A as vertex set and edge set

$$E = \{\{x,y\}: \rho(x,y) \text{ is odd}\}.$$

THEOREM 8: If the integral pseudometric ρ on a set A is a sum of bipartite pseudometrics then the odd distance graph G_ρ of ρ is either the null graph on A or a complete bipartite graph on A.

PROOF: If all $\rho(x,y)$ are even then G_ρ is the null graph (graph with no edges) on A.

Otherwise there is at least one edge $\{a,c\}$ in G_ρ. Let

$$B = \{x: \rho(a,x) \text{ is even}\}$$

$$C = \{x: \rho(a,x) \text{ is odd}\}$$

As $a \in B$ and $c \in C$, both sets are non-empty.

Also they are disjoint and their union is A.

We examine the cases

(i) $x \in B$ and $y \in B$:

From the definition of B, $\rho(a,x)$ and $\rho(a,y)$ are both even. As the triangle $\{a,x,y\}$ cannot have just one odd edge, then $\rho(x,y)$ is even also.

(ii) $x \in C$ and $y \in C$:

As $\rho(a,x)$ and $\rho(a,y)$ are odd, then $\rho(x,y)$ must be even.

(iii) $x \in B$ and $y \in C$:

As $\rho(a,x)$ is even and $\rho(a,y)$ is odd, then $\rho(x,y)$ must be odd.

Thus the edge set E of G_ρ is precisely

$$E = \{\{x,y\}: x \in B \text{ and } y \in C\},$$

and this defines a complete bipartite graph.

This theorem then provides a very easily applied necessary condition for an integral pseudometric to be a sum of bipartite pseudometrics. From our fragmentary state of knowledge this may also be a sufficient condition.

The problem of constructing the various bipartite pseudometric decompositions of an integral pseudometric is more amenable than that of the general decomposition. If A has n members there are $2^n - 1$ bipartite pseudometrics on A. The values of $\rho(x,y)$ over the $\frac{1}{2}n(n-1)$ pairs $\{x,y\}$ with $x \neq y$ mean that the problem is reduced to solving $\frac{1}{2}n(n-1)$ linear diophantine equations in $2^n - 1$ unknowns.

This can be improved on considerably, at least when the values $\rho(x,y)$ are relatively small, by considering the problem in several stages. We restrict ρ first to a set B with three points and then add one point at a time until we reach the whole of A.

This means solving a sequence of problems, but in favourable cases the number of variables is kept small throughout. The method is best indicated by an example.

Consider the set $A = \{1,2,3,4,5\}$ and the integral pseudometric given by the table

```
1
5 │ 2
7   2 │ 3
8   3   1 │ 4
4   3   5   4 │ 5
```

Examination of the odd distances shows that G_ρ is the complete bipartite graph with the sets $B = \{1,4,5\}$ and $C = \{2,3\}$. Thus ρ survives the necessary criterion.

We now take the set $\{1,2,3\}$: any three-part set would do, and there might even be an advantage in taking the set $\{2,3,4\}$ in which the distances are smaller. As the restrictions of the bipartite partitions on $\{1,2,3,4,5\}$ to $\{1,2,3\}$ are either bipartite or trivial, we need to consider first the partitions

$$\{1,2,3\} \qquad x_0$$
$$\{1\}, \{2,3\} \ x_1$$
$$\{2\}, \{1,3\} \ x_2$$
$$\{3\}, \{1,2\} \ x_3$$

where x_0, x_1, x_2, x_3, are the corresponding multiplicities of these partitions.

Matching the distances $\rho(1,2)$, $\rho(1,3)$, $\rho(2,3)$ gives the equations:

$$x_1 + x_2 = \rho(1,2) = 5,$$
$$x_1 + x_3 = \rho(1,3) = 7,$$
$$x_2 + x_3 = \rho(2,3) = 2.$$

We obtain the unique solutions:

$$x_1 = 5, \ x_2 = 0, \ x_3 = 2,$$

but can get no information on x_0.

Thus we have the partial solution:

$$\{1,2,3\} \qquad\qquad \text{unknown}$$
$$\{1\}, \{2,3\} \qquad\quad 5$$
$$\{3\}, \{1,2\} \qquad\quad 2$$

Bringing in point 4, the new point may be added to either subset, so each of the above partitions yields two partitions.

$$\{1,2,3,4\} \qquad y_0$$
$$\{1,2,3\}, \{4\} \ y_1$$
$$\{1,4\}, \{2,3\} \ y_2$$
$$\{1\}\{2,3,4\} \qquad y_3$$
$$\{3,4\}, \{1,2\} \ y_4$$
$$\{3\}, \{1,2,4\} \ y_5$$

Matching the new partitions to those of $\{1,2,3\}$,

$$y_2 + y_3 = 5$$
$$y_4 + y_5 = 2;$$

and from the values of $\rho(1,4)$, $\rho(2,4)$ and $\rho(3,4)$ respectively:

$$y_1 + y_3 + y_4 = 8$$
$$y_1 + y_2 + y_4 = 3$$
$$y_1 + y_2 + y_5 = 1$$

The only solution in non-negative integers yields the following partitions and multiplicities:

$$\{1,2,3,4\} \qquad\quad \text{unknown,}$$
$$\{1,2,3\}, \ \{4\} \qquad 1,$$
$$\{1\}, \{2,3,4\} \qquad 5,$$
$$\{1,2\}, \{3,4\} \qquad 2.$$

From these four sets come seven partitions of $\{1,2,3,4,5\}$ (since A alone is not allowable), three equations linking pairs of new partitions with their parent partition of $\{1,2,3,4\}$, and four equations from the distances $\rho(1,5)$, $\rho(2,5)$, $\rho(3,5)$, $\rho(4,5)$.

It happens that this yields a unique solution:

$$\{1,2,3\}, \{4,5\} \quad 1,$$
$$\{1,5\}, \{2,3,4\} \quad 2,$$
$$\{1\}, \{2,3,4,5\} \quad 3,$$
$$\{1,2,5\}, \{3,4\} \quad 2.$$

In general multiple solutions from earlier stages might need to be followed through.

The equations found in this investigation are of a very restricted type and it seems likely that special methods could much reduce the labour of solving them.

We have thus succeeded in producing a simple necessary condition for the existence of a decomposition into bipartite partitions and a marginally usable method for finding such decompositions. We do not know whether the necessary condition is also sufficient. We believe that the algorithm can be made much more efficient.

As for the existence of a decomposition into general partition pseudometrics, we know little beyond the facts that it is not always possible, and that when it is possible it is often not unique.

REFERENCES

[1] Jardine, N. and Sibson, R. Mathematical Taxonomy. Wiley, New York 1971.

[2] Kelley, J. L. General Topology. Van Nostrand, Princeton, N.J. 1955.

[3] Sokal, R. R. and Sneath P.H.A. Principles of Numerical Taxonomy. W. H. Freeman, San Francisco, 1963.

Department of Mathematics,
University of Canterbury,
Christchurch,
New Zealand.

COMPARISON OF WEIGHTED LABELLED TREES

D. F. ROBINSON and L. R. FOULDS

The results in a previous paper on the comparison of (unweighted) labelled trees are extended to the comparison of weighted labelled trees. An elementary operation is introduced which enables one to transform one weighted labelled tree into another. The operation makes it possible to compare different trees in a quantitative way in the sense of defining a "distance" between them. An application to a problem in molecular evolution is given.

1.   INTRODUCTION

This paper extends the results of a previous paper by Robinson and Foulds (1978) on the comparison of labelled trees to the comparison of weighted labelled trees. The previous results correspond to the special case in which all edges in the trees being compared have unit weight. The approach can be applied to the problem of comparing phylogenetic (evolutionary) trees. This application is discussed later.

DEFINITION 1:   A <u>weighted labelled tree</u> is an ordered quadruple (V,E,S,w) where

       (i) (V,E) is a tree,

      (ii) S is a set of labels which can be partitioned as $\{S_1, S_2, \ldots, S_m\}$ (where $|V| = m$) such that there exists a one-one correspondence

$$\beta: V \rightarrow \{S_1, S_2, \ldots, S_m\}$$

with the property that

$$v \in V, \ d(v) \leqslant 2 \Rightarrow \beta(v) \neq \emptyset$$

(where $d(v)$ is the degree of vertex v).

    (iii) w is a mapping:

$$w: E \rightarrow \mathbf{R}^+, \text{ the positive reals.}$$

The set of all weighted labelled trees with S as the set of labels is denoted by γ_S^w. It is usual to define $\beta(v_i) = S_i$, $i = 1, 2, \ldots, m$, and S_i is called the <u>label</u> of v_i. If $uv \in E$ then $w(uv)$ is called the <u>weight</u> of uv. Examples of weighted labelled trees are given in figure 1.

2.   OPERATION α

If $T_1 = (V_1, E_1, S, w_1) \in \gamma_S^w$ and $v_i v_j \in E_1$, then a tree $T_2 = (V_2, E_2, S, w_2) \in \gamma_S^w$ can be constructed from T_1 by an application of operation α. The operation involves the following steps. A new vertex, $v_{m+1} \notin V_1$ is created where $|V| = m$. Define

$$V_2 = (V_1 \cup \{v_{m+1}\}) \setminus \{v_i, v_j\} .$$

Let

$$E^k = \{v_k v_q : v_k v_q \in E, \; v_q \in V\}, \text{ the set of edges incident with } v_k.$$

Then define

$$E_2 = [(E_1 \setminus E^i) \setminus E^j] \cup \{v_{m+1} v_h : \; v_h v_i \in E^i \text{ or } v_h v_j \in E^j, \; h \neq i \text{ or } j\}.$$

Let

$$\beta(v_h) = S_h \qquad\qquad \forall v_h \in V_2, \; v_h \neq v_i \text{ or } v_j,$$

$$\beta(v_{m+1}) = S_i \cup S_j .$$

Let $w_2(v_k v_q) = w_1(v_k v_q)$, where $v_k v_q \in E_2, v_k \neq v_{m+1}, \; v_q \neq v_{m+1}$,

$$w_2(v_h v_{m+1}) = \begin{cases} w_1(v_h v_i) & \text{if } v_h v_i \in E_1 \\ w_2(v_h v_j) & \text{if } v_h v_j \in E_1 . \end{cases}$$

An example of the application of operation α on a tree in γ_S^w is shown in figure 2.
Operation α can be viewed as a "shrinking" of an edge until its incident points coalesce. The coalesced point is assigned the union of the labels of the two original points. The weights of all other edges in the tree remain the same.

We now state some elementary results concerning the application of operation α to weighted labelled trees. The proofs follow similar lines to those given for unweighted labelled trees in Robinson and Foulds (1978).

DEFINITION 2: The weighted labelled tree $(\{v\}, \emptyset, S, w)$ where $\beta(v) = S$ is a tree with a single vertex, labelled with S, and is denoted by U_S^w.

THEOREM 1: U_S^w can be obtained from any tree $t \in \gamma_S^w$ by a sequence of $(n-1)$ applications of operation α if T has n vertices.

THEOREM 2: If $T \in \gamma_S^w$ and $|S| = n > 1$, then T has at most $(2n-2)$ vertices.

DEFINITION 3: Trees T_1, $T_2 \in \gamma_S^w$ are said to be identical if there exists an isomorphism between them (in the graph-theoretic sense) which preserves labels and weight-identical if weights are also preserved.

Consider a weighted, labelled tree $T = (V, E, S, w) \in \gamma_S^w$ with edge $e \in E$.
The removal of e from T creates two subgraphs T_e' and T_e'' of T which are both weighted labelled trees. Let T_e' and T_e'' have vertex sets V_e' and V_e'' respectively.

Let

$$S_e' = \bigcup_{v \in V_e'} \beta(v)$$

and

$$S_e'' = \bigcup_{v \in V_e''} \beta(v) .$$

Thus S_e' and S_e'' together constitute a partition of S which we denote by $\{S_e', S_e''\}$.
Neither S_e' nor S_e'' can be empty as both T_e' and T_e'' contain vertices which were pendant in T. Thus by definition 1(ii), as pendant vertices are of degree 1, their labels are

nonempty. Let Z_S denote the set of all proper partitions of S into two subsets. Define

$$f : E \to Z_S$$

by

$$f(e) = \{S_e', S_e''\} \qquad \forall e \in E .$$

f is called the <u>partitioning function</u> of T.

THEOREM 3: f is one-to-one.

COROLLARY

The number of trees obtainable from a given tree, $T \in \gamma_S^w$, which has n vertices, by a single application of operation α is (n-1).

DEFINITION 4: Let T_1, $T_2 \in \gamma_S^w$ with edge sets E_1 and E_2 and partitioning functions f_1 and f_2 respectively. Then edges $e_1 \in E_1$ and $e_2 \in E_2$ are said to be <u>matched</u> if and only if

$$f_1(e_1) = f_2(e_2).$$

There will be a one-to-one correspondence between matched edges in T_1 and T_2. We now define

$$E_1' = \{e_1: \ e_1 \in E \ , \ \exists e_2 \in E_2 \ \text{s.t.} \ f_1(e_1) = f_2(e_2)\}$$
$$E_2' = \{e_2: \ e_2 \in E_2, \ \exists e_1 \in E_1 \ \text{s.t.} \ f_2(e_2) = f_1(e_1)\}$$

i.e. E_1' and E_2' comprise the edges in T_1 and T_2 respectively which can be matched.

THEOREM 4: T_1, $T_2 \in \gamma_S^w$ with edge sets E_1 and E_2 respectively are weight identical
$\Leftrightarrow$ there exists a one-to-one correspondence h: $E_1 \to E_2$ such that

$$e \in E_1 \Rightarrow \text{(i) e and h(e) are matched}$$
$$\text{(ii)} \ w(e) = w(h(e)).$$

3.   <u>DISTANCE BETWEEN TREES IN γ_S^w</u>

We come now to comparing trees in γ_S^w with a view to defining a "distance" between them. It is desirable (in terms of the application to phylogeny which follows) that this distance should have the following properties:

(i)   It should be easily calculated for any pair of trees in γ_S^w.

(ii) It should be represented by a nonnegative integer.

(iii) Two trees with similar topologies and equal matched edge weights should be a smaller distance apart than two trees with radically different topologies and matched edges with different weights.

Consider two trees, T_1, $T_2 \in \gamma_S^w$ with edge sets E_1 and E_2 respectively. On comparing them E_1' and E_2' can be formed, which may possibly be empty. Consider the trees T_1' and T_2' obtained from T_1 and T_2 respectively by removing all edges in $E_1 \setminus E_1'$ and $E_2 \setminus E_2'$ by operation α in any order. T_1' and T_2' will be identical.

Indeed if h : $E_1 \to E_2$ is the matching function for T_1 and T_2, and

$$w(e) = w(h(e)) \qquad\qquad \forall e \in E_1'$$

then T_1' and T_2' are weight identical.

DEFINITION 5: Let T_1, $T_2 \in \gamma_S^w$, with edge sets E_1 and E_2 respectively. The <u>distance</u> between T_1 and T_2,

$$d(T_1, T_2) = \sum_{e \in E_1 \setminus E_1'} w(e) \quad + \sum_{e \in E_1'} |w(e) - w(h(e))| + \sum_{e \in E_2 \setminus E_2'} w(e).$$

Some results concerning d are now stated.

THEOREM 5: For all T_1, $T_2 \in \gamma_S^w$

    (i) $d(T_1, T_2) = d(T_1, T_1') + d(T_1', T_2'), + d(T_2', T_2)$

    (ii)   If $E_i' = E_i$ (i.e. every edge in T_1 is matched with an edge in E_2 and vice versa) then T_1 and T_2 are identical and $T_i' = T_i$, $i = 1,2$. Thus

$$d(T_1, T_2) = \sum_{e \in E_1} |w(e) - w(h(e))|$$

    (iii) If $E_1' = \emptyset$, (i.e. there are no matched edges), then

$$T_1' = T_2' = U_S^w$$

    and

$$d(T_1, T_2) = d(T_1, U_S^w) + d(U_S^w, T) = \sum_{e \in E_1 \cup E_2} w(e).$$

The application of d to calculating distances between weighted labelled trees is given in the next section.

4.   <u>AN APPLICATION TO MOLECULAR EVOLUTION.</u>

The theory of evolution asserts that existing biological species have been linked in the past by common ancestors. A diagram showing these links is a phylogeny or phylogenetic tree. There has been no general agreement on methods of determining phylogenies, particularly when there is no fossil record and it is not surprising that there is often little agreement even on the main outline of evolution within some major classes. Several attempts have been made to establish more objective methods of determining phylogenies. Some workers have constructed phylogenies by considering the mutations of a selected protein which is **possessed by all species in the class under** study. In particular we shall consider the case where the data are amino acid sequences, one sequence for each species. The sequences are of the same length and thus alignable such that the kth amino acid character (position) for all species may be assumed to have descended from the same ancestral amino acid. Each amino acid is associated with ordered triples of nucleic acid bases. Given any two acids it is possible to determine how to transform one into the other in the minimum number of base changes. Hence given any two sequences, representing different species, it is

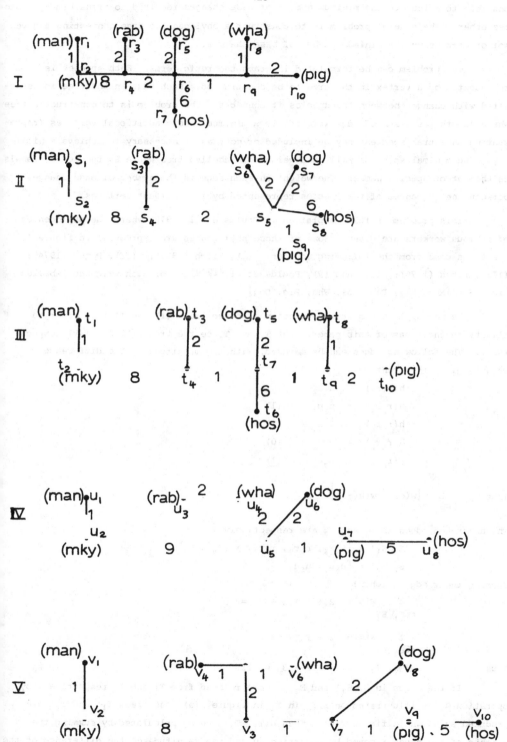

Fig. 1 Phylogenetic Trees for Seven Mammalian Species

possible to calculate the minimum number of base changes required to transform one into the other. The general problem is to construct a phylogenetic tree connecting a given set of species by the minimum number of base changes.

The problem can be formulated in graph theoretic terms. Each species is represented by a vertex in the tree to be constructed. Each edge in the tree is associated with changes between the species it connects. The problem is to construct a tree whose length (as a sum of edge weights) is a minimum, where additional vertices (representing ancestral species) may be included or not, as is necessary to achieve minimality. The minimal solution will be a weighted, labelled tree where S, the set of labels is the set of species names. The weight on each edge is the number of base changes between the sequences of the species represented by its incident vertices.

This problem is fully explained in Foulds et al (1978) where the phylogenies of various workers are given. Some of those phylogenies are reproduced in figure 1. The trees come from the following sources: (I), Fitch (]973); (II), Penny (1974); (III), Fitch (1976); (IV) and (V), Foulds et. al. (1979). For each weighted labelled tree, S = {Man, Mky, Rab, Hos, Wha, Pig, Dog}.

We now apply the distance function d to two trees (I) and (IV) in order to illustrate the ideas of this paper. Let T_1 and T_2 be the trees (I) and (IV) respectively. The following edges can be matched, with the magnitude of the difference of weights given in parenthesis.

$$h(r_1 r_2) = u_1 u_2 \qquad (0)$$
$$h(r_2 r_4) = u_2 u_5 \qquad (1)$$
$$h(r_3 r_4) = u_3 u_4 \qquad (0)$$
$$h(r_5 r_6) = u_5 u_6 \qquad (0)$$
$$h(r_6 r_7) = u_7 u_8 \qquad \underline{(1)}$$
$$2$$

Thus $\displaystyle\sum_{e \in E_1'} |w(e) - w(h(e))| = 2$

The unmatched edges in T_1 and T_2 are respectively:

$$E_1 \backslash E_1' = \{r_4 r_6, r_6 r_9, r_8 r_9, r_9 r_{10}\},$$
$$E_2 \backslash E_2' = \{u_4 u_5, u_5 u_7\}.$$

Summing their edge weights:

$$\sum_{e \in E_1 \backslash E_1'} w(e) = 2 + 1 + 1 + 2 = 6$$

$$\sum_{e \in E_2 \backslash E_2'} w(e) = 2 + 1 = 3$$

Thus $\qquad d(T_1, T_2) = 6 + 2 + 3 = 11.$

If the edges in $E_1 \backslash E_1'$ and $E_2 \backslash E_2'$ are removed from T_1 and T_2 respectively by operation α, as illustrated on $r_4 r_6$ in T_1 in figure 2(a), the trees T_1' and T_2' are produced as shown in figure 2(b). Then $d(T_1, T_2)$ can be calculated by summing the weights of the edges removed by operation α and the magnitude of the difference of the matched weights in T_1' and T_2'.

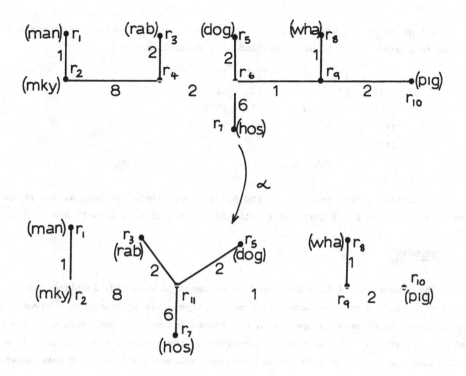

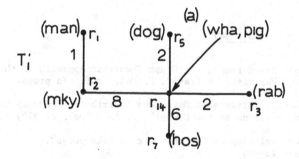

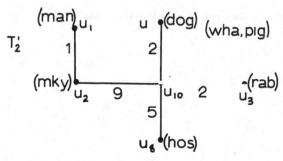

Fig. 2 The Comparison of Two Phylogenetic Trees

All pairs of the trees in figure 1 have been compared using the method out-
lined in this paper. The table of distances is given in table 1.

	(II)	(III)	(IV)	(V)
(I)	3	1	11	11
(II)		4	9	10
(III)			9	10
(IV)				4

Table 1.

This table indicates that trees I, II and III are relatively similar, as are IV and V.
However, each of the first three is relatively quite different from IV and V.

5.   SUMMARY

The comparison of labelled trees is extended to weighted, labelled trees. An
elementary operation is introduced which makes it possible to transform a given tree
into another. Based on this operation, a distance between any two trees with the same
label set is defined. The problem of constructing phylogenetic trees is introduced.
The distance concept is used to compare various phylogenetic trees that have appeared
in the literature.

REFERENCES

[1] L.R. Foulds, M.D. Hendy and David Penny, "A Graph Theoretic Approach to the
     Development of Minimal Phylogenetic Trees", J. Mol. Evol. In press.

[2] W.M. Fitch, "Is the Fixation of Observable Mutations Distributed Randomly among
     the Three Nucleotide Positions of the Codon?", J. Mol. Evol. 2(1973) 123-136.

[3] W.M. Fitch, "The Molecular Evolution of Cytochrome c in Eukaryotes".
     J. Mol. Evol. 8(1976) 13-40.

[4] David Penny, "Evolutionary Clock: The Rate of Evolution of Rattlesnake
     Cytochrome c", J. Mol. Evol. 3(1974) 179-188.

[5] D.F. Robinson and L.R. Foulds", Comparison of Labelled Trees",
     J. Comb. Th. (submitted).

Mathematics Department,
University of Canterbury,
Christchurch, N.Z.

Mathematics Department,
Massey University,
Palmerston North N.Z.

ISOMORPHIC FACTORISATIONS VI: AUTOMORPHISMS

R.W. ROBINSON

An isomorphic factorisation *of a graph* G *is a partition of its line set* E(G) *into isomorphic subgraphs called* factor *graphs. The question investigated here is how those automorphisms of a complete graph* K_p *which preserve an isomorphic factorisation can act in permuting the factor graphs. The group of these permutations is called the* symmetry group. *It is clear that the symmetric group* S_2 *is the only such factor group of degree 2. However, it is shown that all four permutation groups of degree 3 arise from isomorphic factorisations of complete graphs. For* $E_1 \times S_2$ *the smallest example requires 10 points. A natural conjecture is that every permutation group of degree d > 2 is the symmetry group of an isomorphic factorisation of some complete graph. However no such representation is known for* S_6, *and it is shown that the representations of* S_5 *present certain irregularities.*

1. INTRODUCTION

Given a graph G = (V,E), a *factorisation* of G is a partition P = $(E_1,\ldots,E_t)$ of the line set E. If $(V,E_i) \cong (V,E_j)$ for all $1 \le i$, $j \le t$, then P is an *isomorphic factorisation* of G. Each subgraph (V,E_i), $1 \le i \le t$, is a *factor graph* of P. The *full symmetry group* of the factorisation is the set of permutations on V which preserve the partition P. The *symmetry group* of the factorisation is the action of the full symmetry group on the members $E_1,\ldots,E_t$ of the partition. The kernel of the canonical mapping from the full symmetry group onto the symmetry group is called the *strict symmetry group* of P. It consists of the permutations of V which fix $E_1,E_2,\ldots,$ and E_t individually.

In this paper we study the question of which finite permutation groups can be obtained as symmetry groups of isomorphic factorisations of complete graphs. As usual S_n, Z_n and I_n denote the symmetric, cyclic and identity groups, respectively, of degree n. It is shown by example in the next two sections that I_3, $I_1 \times S_2$, Z_3, S_3, S_4 and S_5 can all be represented as symmetry groups of isomorphic factorisations of complete graphs. The only obvious example of a permutation group which cannot be so represented is I_2. It is unknown whether every other finite permutation group can be so represented.

In the final section some related results and questions are presented. Part of the extensive literature on 1-factorisations of complete graphs is summarised. A 1-factor is a subgraph in which every point has degree 1, that is, a point-disjoint union of copies of K_2. A 1-factorisation is a partition into line-disjoint 1-factors, and is therefore a special sort of isomorphic factorisation.

Isomorphic factorisations were introduced in full generality in [10]. There it was shown that if $t \mid \binom{p}{2}$, then the complete graph K_p has an isomorphic factorisation into t factor graphs. Thus the divisibility condition, that t divides the total number of lines, was shown in this case to be sufficient to ensure divisibility of complete graphs. The sufficiency of the divisibility condition for complete multipartite graphs is investigated in [11] and failures were found for complete tripartite graphs and odd t, and for complete tetrapartite graphs and $t = 2$. Similar results are obtained for complete digraphs and complete multipartite digraphs in [12]. The correspondence between isomorphic factorisations of graphs and various sorts of combinatorial designs is expounded in [13]. An analogue of Ramsey numbers based on isomorphic factorisations is introduced in [9].

For standard graph theoretic terminology we follow the book by Harary [8].

2. GROUPS OF DEGREE 3

In this section we show that all four permutation groups of degree 3 are obtainable as symmetry groups of isomorphic factorisations of complete graphs. By contrast, the identity group of degree 2 cannot be represented in such a fashion. For an isomorphic factorisation of K_p into two factors consists of a self-complementary graph G on p points, together with its complement $\overline{G}$. Any isomorphism of G to $\overline{G}$ perforce maps $\overline{G}$ to G, since all the lines of K_p lie in one or the other. Thus the symmetry group of $\{G, \overline{G}\}$ must be S_2. It is well known that self-complementary graphs on p points exist whenever allowed by the divisibility condition, that is, whenever $p \equiv 0$ or $1 \pmod 4$. In particular this is a consequence of the main theorem of [10].

Since the number of lines in K_p is $p(p-1)/2$, which is divisible by 3 only if $p \equiv 0$ or $1 \pmod 3$, there can be no isomorphic factorisation of K_p into three factors unless $p = 3n$ or $p = 3n + 1$. For $p = 3$, there is only one factorisation of K_3 into three parts and its symmetry group is S_3. This generalises to a factorisation of K_{3n} into three copies of $K_n \cup K_{n,n}$ illustrated in Figure 1, again with symmetry group S_3. One then obtains a factorisation of K_{3n+1} with the same symmetry group by joining a new point to the K_n in each factor. In the figure, each circle represents K_n if closed or $\overline{K_n}$ if open, while each line represents the join of the point sets adjacent to it.

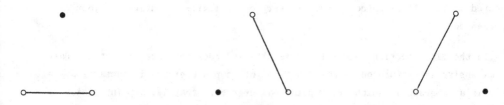

FIGURE 1 - A factorisation of K_{3n} with symmetry group S_3

Isomorphic factorisations of K_{16} with S_3 as symmetry group occur naturally in the study of Ramsey numbers. To show that $R_3(3,2) > 16$, one requires a proper 3-colouring of the lines of K_{16}, that is, a colouring with no monochromatic triangles. It turns out that there are just two of these, each having symmetry group S_3 on the colours. Both are based on a single factor graph. The full symmetry groups of these two factorisations have orders 960 and 192, as computed in [16,pp.40-43].

For $p = 4$, there are two isomorphic factorisations of K_4 into three parts. One is the unique 1-factorisation, with symmetry group S_3. The other has factors isomorphic to $K_1 \cup K_{1,2}$ and has Z_3 as its symmetry group. For $n > 1$ this can be generalised to a factorisation of K_{3n} into copies of $\overline{K_n} \cup (K_n + \overline{K_n})$ as shown in Figure 2 with Z_3 still the symmetry group. As before one obtains a factorisation of K_{3n+1} by joining a new point to the K_n in each factor. In this case Z_3 is the symmetry group for $n \geqslant 1$.

FIGURE 2 - A factorisation of K_{3n} with symmetry group Z_3, $n > 1$

The smallest example of an isomorphic factorisation of a complete graph with I_3 as symmetry group was originally discovered by J. Schoenheim (unpublished). It is a factorisation of K_6 into copies of the path P_5. One can show by exhaustion that all other isomorphic factorisations of K_6 into 3 parts have symmetry group Z_3 or S_3. In fact P_5 is unique in appearing in more than one such factorisation of K_6. The other factorisation for P_5 has symmetry group Z_3. Schoenheim's factorisation is the $m = 0$ case of the scheme represented in Figure 3. The partly closed circles represent the factors of an isomorphic factorisation of K_m into three parts, $m = 3n$ or $3n + 1$ and $n \geqslant 0$.

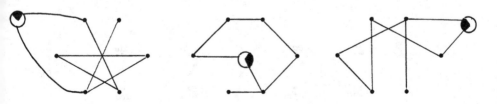

FIGURE 3 - An isomorphic factorisation of K_p with symmetry group I_3, $p = 3n + 6$ or $3n + 7$ and $n \geqslant 0$

For each m considered, at least one such factorisation exists, by the main theorem of [10]. Whichever factorisation is chosen, the result is an isomorphic factorisation of K_{m+6} with symmetry group I_3.

It has already been remarked that no isomorphic factorisation of K_p with symmetry group $I_1 \times S_2$ exists for $p \leqslant 6$. The same is true for $p = 7$ and, apparently, for $p = 9$. (The author's exhaustive search in the latter case was not careful enough to warrant certainty.) In Figure 4 an example is shown for K_{10}. It is not clear how to generalise this to $p = 3m + 10$ and $3m + 12$, though it seems likely that isomorphic factorisations of K_p with symmetry group $I_1 \times S_2$ exist for all $m \geqslant 0$. In the figure the three factors are drawn separately, with the six point orbits labelled a through f in order to aid recognition of isomorphisms between them. The first factor is invariant under the permutation fixing points a and f and interchanging those pairs labelled b,c,d and e. This permutation is seen to interchange the second and third factors. At the bottom, each point of K_{10} is labelled with the orbits it belongs to in the three factors. The pattern is obviously inconsistent with an automorphism mapping the factors to each other in a 3-cycle.

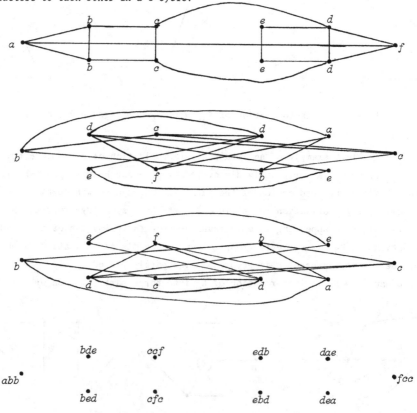

FIGURE 4 - An isomorphic factorisation of K_{10} with symmetry group $I_1 \times S_2$

3. SMALL SYMMETRIC GROUPS

In this section we consider the existence of isomorphic factorisations of K_p with symmetry group S_k, for $k \leqslant 6$. As stated in the proposition below, a necessary condition is that $p \equiv 0$ or $1 \pmod k$ for odd k and $p \equiv 0$ or $1 \pmod{2k}$ for even k. In the previous section it was shown that for $k = 2$ and 3 this condition is also sufficient. We will show that this condition is sufficient for $k = 4$, but not for $k = 5$ or 6. For $k = 5$ an infinite family of factorisations with symmetry group S_5 is exhibited. For $k \geqslant 6$ it is not known whether any factorisation with S_k as symmetry group exists.

Proposition 1.   *If a factorisation of* K_p *into k factors admits an automorphism which permutes the factors in a k-cycle, then* $p \equiv 0$ *or* 1 *(mod k) if k is odd and* $p \equiv 0$ *or* 1 *(mod 2k) if k is even.*

Proof.   The hypothesised automorphism of K_p must permute the lines in cycles of lengths all divisible by k. For k odd, this requires that the points all be permuted in cycles of lengths divisible by k with the exception of at most one fixed point. Thus $p = mk$ or $mk + 1$ for some m. For an even length point cycle, the induced diagonal line cycle has half its length. Thus when k is even the point cycle lengths must all be divisible by $2k$, again with the exception of at most one fixed point. So $p = 2mk$ or $2mk + 1$ for some m.

For $k = 4$, the affine geometry of dimension 2 over GF(3) provides an isomorphic factorisation of K_9 into four copies of $3K_3$ with symmetry group S_4. In general, the affine geometry of dimension d over GF(q) can be constructed on a d-dimensional vector space V over GF(q) as point set, the lines being all cosets of the rank 1 subspaces of V. To obtain a factorisation of the complete graph on V, place two point pairs of V in the same class of the partition just if the geometric lines they determine are parallel. Thus there are q^d points in all and $(q^d-1)/(q-1)$ factors, one for each rank 1 subspace of V. The graphical line joining two points a,b in V is a member of the factor corresponding to the subspace generated by $a - b$. The automorphisms of these affine geometries are well understood; see [5,p.32]. If $q = p^e$ for prime p, the strict automorphism group has order $q^d(q-1)e$. This is generated by the q^d translations given by the additive group of V, the $q - 1$ dilations given by the multiplicative group of GF(q), and the e automorphisms of GF(q). The symmetry group is given by the action of the projective general linear group PGL(d,q) on the rank 1 subspaces of V. The order of the symmetry group is therefore $(q^d-1)\cdot\ldots\cdot(q^d-q^{d-1})/(q-1)$. When $d = 2$ and $q = 3$ there are 4 factors and 24 automorphisms of them, so the symmetry group must be S_4. In general each factor is a copy of $q^{d-1}K_q$, so in this case we have 4 copies of $3K_3$.

The action of PGL(d,q) leaves the origin of V fixed. Taking away the origin therefore gives an isomorphic factorisation of the complete graph on the remaining $q^d - 1$ points. The symmetry group is still $PGL(d,q)$, although the strict automorphism group is reduced to order $(q-1)e$ since the nontrivial translations are now omitted. When $d = 2$ and $q = 3$ this provides a factorisation of K_8 into four copies of $K_2 \cup 2K_3$ with S_4 as symmetry group. Moreover we can replace the missing origin by any isomorphic factorisation P of K_{8m} or K_{8m+1} having symmetry group S_4. The four factors of P can be identified in any order with the existing four factors of K_8. Any line from a point of P to a point b of the existing K_8 is assigned to the same factor as the line joining the origin to b in our factorisation of K_9. The symmetry group of the resulting factorisation of K_{8m+8} or K_{8m+9} is clearly S_4 again. Thus by induction we see that an isomorphic factorisation of K_p into four parts with S_4 as symmetry group exists whenever $p \equiv 0$ or $1 \pmod 8$.

It is not hard to verify that up to isomorphism there are just two factorisations of K_8 with symmetry group S_4. The one that does not arise from an affine geometry has factors isomorphic to $K_2 \cup C_6$. It can be constructed by starting with $(15)(238674)$, where the digits represent points and cyclically adjacent digits are to be adjacent in this factor. Then the other three factors are represented by $(26)(341765)$, $(37)(452876)$ and $(48)(563187)$. It is clear from the construction that the point permutation $i \mapsto i + 1 \pmod 8$ is an automorphism which maps the factors in a 4-cycle. The point permutation $(1)(5)(278)(346)$ fixes the first factor and permutes the others in a 3-cycle. Thus the full symmetry group of the factorisation must be S_4. As before one can extend this by adding any factorisation P of a complete graph with symmetry group S_4. Lines from points in P to i for $1 \leqslant i \leqslant 8$ are added to that factor of K_8 in which i has degree 1.

For $k = 5$, the unique 1-factorisation of K_6 has symmetry group S_5. This is well known; see [3,Lemma 3] or [4,Theorem 4.7]. Let P be a particular example of this factorisation on K_6, with the five line classes labelled 1 through 5. A factorisation of K_{30} with S_5 as symmetry group can be constructed from P by replacing each point a by a copy of $P - a$. The lines within $P - a$ are assigned the same factors as they were in P, while all lines from $P - a$ to $P - b$ are assigned to the same factor as the line $\{a,b\}$ in P. It is even easier to construct a factorisation of K_{36} with S_5 as symmetry group since each point of P can be replaced by a copy of P itself. These constructions can clearly be extended to give factorisations of K_p having symmetry group S_5 for infinitely many values of p. However, for $p \leqslant 36$, the examples given for $p = 6$, 30 and 36 are the only ones known. It is quite easy to verify that none exists for $p = 5$. The next proposition shows that none exists for $p = 10$, and the same methods suffice just as easily to show that none exists for $p = 11$.

Proposition 2. *There is no isomorphic factorisation of K_{10} into five factors having S_5 as its symmetry group.*

Proof.   Suppose to the contrary that P is a factorisation of K_{10} into
five copies $G_1, \ldots, G_5$ of a graph G, and that P has symmetry group S_5 on these factors.
Then P has an automorphism α which fixes one factor and permutes the others in a 4-
cycle; we may assume that G_1 is the factor left fixed. Now α is an automorphism of
G_1 in which every line of the complement $\overline{G}_1$ is contained in a line cycle of length
divisible by 4. Considering that G_1 must have 9 lines and 10 points, it is evident
that α must consist of either an 8-cycle or two 4-cycles, along with either a
transposition or two fixed points. The graphs with 9 lines left fixed by such a
point permutation are not hard to list. Taking first the permutations containing an
8-cycle, one finds that the only possibilities are $K_{1,9}$, $K_2 \cup C_8$, $K_2 \cup 2C_4$, and the
tree T shown in Figure 5. The two disconnected graphs which are also shown in
Figure 5 are the only additional possibilities obtained when the point permutation
contains two 4-cycles. So G must be isomorphic to one of these six graphs.

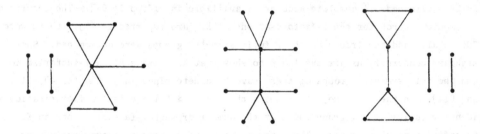

FIGURE 5- Three possible factors of K_{10}

In addition, P must contain an automorphism β which interchanges two of the
factors, say G_1 and G_2, and permutes the others in a 3-cycle. Then β^2 is an automorphism of G_1 which has order divisible by 3, since it still permutes G_3, G_4 and G_5
in a 3-cycle. Of the six possibilities for G, only $K_{1,9}$ and T have automorphisms
of order 3. Of course, $K_{1,9}$ cannot appear in any isomorphic factorisation since
its complement contains an isolated point, so only T need by considered.

Now the automorphism β of P must permute all lines in cycles of length divisible by 2 or 3, with at least 18 lines in cycles of length divisible by 2 and 27
lines in cycles of length divisible by 3. Any point permutation satisfying these conditions has one of the following three patterns of point cycle lengths: one each
of lengths 1, 3 and 6; one of length 4 and two of length 3; one each of lengths
4 and 6. In the first case β^2 has a unique fixed point, which is impossible for an
automorphism of T since the central line must be preserved (possibly with endpoints
reversed). In the other two cases the line cycles of length 2 and 4 induced by the
4-cycle of points must alternate lines in G_1 with lines in G_2, so G_1 contains a
path of length 3 formed by opposite sides of a square and one of the diagonals in

the point 4-cycle. The other six lines of G_1 can only lie in one of the two 12-cycles of lines joining the point 4-cycle to the two 3-cycles of points, or to the 6-cycle of points as the case may be. In either situation three points would have to be isolates in G_1, contradicting the fact that it must be isomorphic to the tree T.

The same sort of methods can be applied to show that no isomorphic factorisation of K_{12} or K_{13} can have S_6 as its symmetry group. Thus, by Proposition 1, if K_p has a factorisation with symmetry group S_6 then $p \geq 24$. We do not know whether such a factorisation exists.

4. RELATED RESULTS AND QUESTIONS

There has been considerable interest in the 1-factorisations of K_{2n}. There is only one for each of the values $n = 1, 2, 3$, and for $n = 4$ there are six. This was proved some time ago in [6] where the full symmetry groups of the six were almost fully determined. A complete account is available in [17, pp.91-94]. The results of a computer search for the 1-factorisations of K_{10} are reported in [7]; there were 396 in all, and the orders of their full symmetry groups were determined. Some standard constructions are available to show that K_{2n} always has a 1-factorisation, and the full symmetry groups of these have been determined in [4, Chapter 4], [1], [2] and [14]. In the last two, it is shown that for $n \geq 5$ there is a 1-factorisation with identity full symmetry group and that the number of such 1-factorisations in K_{2n} goes to infinity as n increases. These are examples of isomorphic factorisations of complete graphs with symmetry group I_{2n-1} for $n \geq 5$.

The general question of which permutation groups occur as symmetry groups of isomorphic factorisations of complete graphs appears to be difficult. One could also ask, for each finite group Γ, which complete graphs K_p admit such a factorisation. This seems to present points of interest already when the group is S_5. For, as discussed in the previous section, the obvious necessary condition $p \equiv 0$ or 1 (mod 5) is not always sufficient, with factorisations of the required sort existing when $p = 6, 30, 36$, and for infinitely many higher values of p, but not when $p = 5, 10$ or 11.

For abstract groups the outlook is more hopeful, and we believe that the following is true.

Conjecture. *Every finite group is isomorphic to the symmetry group of some isomorphic factorisation of a complete graph.*

The following proposition goes part way towards proving the conjecture. It is a direct consequence of the fact that any finite group is isomorphic to a subgroup of PGL(d,q) for all sufficiently large d. We saw in the last section that these are all obtained as symmetry groups of affine geometries, which can be viewed as isomorphic factorisations of complete graphs.

Proposition 3. *Every finite group is isomorphic to a subgroup of the symmetry group of some isomorphic factorisation of a complete graph.*

Sharper results along these lines have been obtained in the case of finding a representation as a subgroup of the full symmetry group of an isomorphic factorisation where this subgroup acts regularly on the factors. In [18,Theorem 2] it is shown for an abelian group of order m that a necessary and sufficient condition for the existence of such a representation in a factorisation of K_p, $p \geq 2$, is that m be odd and $p \equiv 0$ or 1 (mod m). A similar result is obtained in [15] for any finite group.

REFERENCES

[1]   B.A. Anderson, Symmetry groups of some perfect 1-factorizations of complete graphs, *Discrete Math.*, 18 (1977), 227-234.

[2]   B.A. Anderson, M.M. Barge and D. Morse, A recursive construction of asymmetric 1-factorizations, *Aequationes Math.*, 15 (1977), 201-211.

[3]   P.J. Cameron, On groups of degree n and n - 1, and highly-symmetric edge colourings, *J. London Math. Soc.* (2), 9 (1975), 385-391.

[4]   P.J. Cameron, *Parallelisms of Complete Designs.* (Cambridge Univ. Press, 1976).

[5]   P. Dembowski, *Finite Geometries.* (Springer-Verlag, New York, 1968).

[6]   L.E. Dickson and F.H. Safford, Solution to problem 8 (group theory), *Amer. Math. Monthly*, 13 (1906), 150-151.

[7]   E.N. Gelling and R.E. Odeh, On 1-factorizations of the complete graph and relationships to round robin schedules, *Proc. 3rd Manitoba Conf. on Numerical Math.* (Utilitas, Winnipeg, 1974), 213-221.

[8]   F. Harary, *Graph Theory.* (Addison-Wesley, Reading, Mass., 1969).

[9]   F. Harary and R.W. Robinson, Generalised Ramsey theory IX: isomorphic factorizations IV: isomorphic Ramsey numbers, *Pacific J. Math.*, to appear.

[10]   F. Harary, R.W. Robinson and N.C. Wormald, Isomorphic factorisations I: complete graphs, *Trans. Amer. Math. Soc.*, 242 (1978), 243-260.

[11]   F. Harary, R.W. Robinson and N.C. Wormald, Isomorphic factorisations III: complete multipartite graphs, *Combinatorial Mathematics,* Proceedings of the International Conference on Combinatorial Theory (Canberra), (Lecture Notes No. 686, Springer-Verlag, Berlin, 1978), 47-54.

[12]   F. Harary, R.W. Robinson and N.C. Wormald, Isomorphic factorisations V: directed graphs, *Mathematika*, to appear.

[13]   F. Harary and W.D. Wallis, Isomorphic factorizations II: combinatorial designs, *Proc. 8th Southeastern Conf. on Combinatorics, Graph Theory and Computing.* (Utilitas, Winnipeg, 1977), 13-28.

[14]   C.C. Lindner, E. Mendelsohn and A. Rosa, On the number of 1-factorizations of the complete graph, *J. Combinatorial Theory (B),* 20 (1976), 265-282.

[15]   Š. Porubský, Factorial regular representation of groups in complete graphs, to appear.

[16]   A.P. Street and W.D. Wallis, Sum-free sets, coloured graphs and designs, *J. Austral. Math. Soc.*, 22(A) (1976), 35-53.

[17]   W.D. Wallis, A.P. Street and J.S. Wallis, *Combinatorics: Room Squares, Sum-free Sets, Hadamard Matrices*. (Lecture Notes No. 292, Springer-Verlag, Berlin, 1972).

[18]   B. Zelinka, Decomposition of the complete graph according to a given group, *Mat. Časopis Sloven. Akad. Vied*, 17 (1967), 234-239. (Czech, with English summary).

A FAMILY OF WEAKLY SELF-DUAL CODES

CHRISTOPHER A. RODGER

This note looks at a family of weakly self-dual codes, C, over GF(3), with (m×2m) generating matrix

$$G = [J-I \ A] \ ,$$

where J is the (m×m) matrix with every entry being +1, I the (m×m) identity matrix, and the circulant incidence matrix of the (4t-1,2t-1,t-1) block design, with first row being formed from the quadratic residues, mod(4t-1) (so m = 4t-1, and is prime).

In particular, we consider these codes for t ≡ 2 (mod 3) which is a necessary and sufficient condition for

$$GG^T = 0.$$

In this case, we show C is contained within the code generated by

$$F = [I \ H]$$

where H is the [(m+1) × (m+1)] Hadamard matrix. However, it is only by careful reduction of the length and dimension of F that allows us to reach a doubly circulant generating matrix.

We show the upper bound for the minimum distance for these codes is 2t + 2. The minimum distance, d, for the first two cases of this family of (8t-2,4t-2,d) weakly self-dual codes are evaluated, showing the upper bound is in fact attained, i.e., there are (14,6,6) and (38,18,12) weakly self-dual codes over GF(3).

An efficient means for finding the minimum weight of this family is given.

INTRODUCTION AND DEFINITIONS

A (n,k,d) linear code C over GF(3) consists of 3^k vectors, called codewords, each of length n, with the minimum distance between codewords being d, and for which $\underline{u}, \underline{v} \in C$ implies that $\underline{u} + \underline{v} \in C$ and $a\underline{u} \ C, \ \forall \ a \in GF(3)$. Thus, all the codewords of C (the family of weakly self-dual codes with generating matrix G) may be found by forming all the linear combinations of the rows of G over GF(3).

The dot product is defined by:

$$\underline{u} \cdot \underline{v} = \sum_{i=1}^{n} u_i v_i$$

with the sum evaluated in GF(3). Addition and subtraction are defined component-wise, so that, for $\underline{u}, \underline{v} \in C$,

$$\underset{\sim}{u} + \underset{\sim}{v} = (u_1 + v_1, u_2 + v_2, \ldots, u_n + v_n)$$

$$\underset{\sim}{u} - \underset{\sim}{v} = (u_1 - v_1, u_2 - v_2, \ldots, u_n - v_n).$$

The dual code $C^{\perp}$ is defined by:

$$C^{\perp} = \{\underset{\sim}{v} \mid \underset{\sim}{u} . \underset{\sim}{v} = 0, \quad u \in C\}.$$

A code is weakly self-dual if $C \subset C^{\perp}$. (v,k,λ) block designs and quadratic residues are defined in [1].

Let $g_i = (\underset{\sim}{x}_i, \underset{\sim}{z}_i)$ denote the ith row of G, where both $\underset{\sim}{x}_i$ and $\underset{\sim}{z}_i$ are of length m, corresponding to the two distinct parts of the generating matrix.

UPPER BOUND FOR THE MINIMUM DISTANCE

Consider the weight of the vector resulting from subtracting the jth row from the ith row at G. Clearly,

$$\text{weight}(\underset{\sim}{x}_i - \underset{\sim}{x}_j) = 2.$$

Now, as $\underset{\sim}{z}_i$ and $\underset{\sim}{z}_j$ are rows of a $(4t-1, 2t-1, t-1)$ block design, $\underset{\sim}{z}_i - \underset{\sim}{z}_j$ will be zero in the t components where $\underset{\sim}{z}_i$ and $\underset{\sim}{z}_j$ have a 0 in common and in the $t-1$ places they have a 1 in common. This leave 2t non-zero components.

Hence,

$$\text{weight}(z_i - z_j) = 2t,$$
$$\text{minimum weight of this code} \leq 2t + 2,$$
$$\text{and minimum distance of this code} \leq 2t + 2.$$

CONDITION FOR $GG^T = 0$

In general, for a (v,k,λ) block design with incidence matrix B:

$$BB^T = (k-\lambda)I + \lambda J.$$

So for A:

$$AA^T = tI + (t-1)J.$$

Also, as $m = 4t - 1$,

$$[J-I] [J-I]^T = (4t-3)J + I.$$

Therefore, for $GG^T = 0$ over $GF(3)$, we require:

$$[J-I] [J-I]^T + AA^T = 0.$$

Equating the coefficients of J and I to zero yields:

$$4t - 3 + t - 1 \equiv 0 \mod (3)$$
$$t + 1 \equiv 0 \mod (3)$$
$$\therefore \quad t \equiv 2 \mod (3).$$

As the rank of C is $4t - 2$ (see next section), C is not self-dual, but $C \subset C^\perp$, nd so is weakly self-dual.

COMPUTING THE MINIMUM WEIGHT

Searching through all the codewords to find the minimum weight is out of the uestion, as there are 3^{m-1} of them. So, we present 3 ways to cut this to a practicable umber.

1.   If we could place an ordering on the codewords, so that we could make a isting of all the codewords less than a certain weight, then we would only need to est the codewords up to the weight $2t + 1$ (as we know there are codewords of weight $t + 2$).

Consider the case where $t \equiv 2 \pmod 3$. Let $\underset{\sim}{Y}$ be a linear combination of ows of $[J-I]$. We wish to find $\underset{\sim}{x}$, such that

$$\underset{\sim}{x}[J-I] = \underset{\sim}{Y} \pmod 3,$$

.e., given $\underset{\sim}{Y}$, what combination $\underset{\sim}{x}$ of rows of $[J-I]$ give rise to it. Once this is done, e can choose all the $\underset{\sim}{Y}$ of weight $<2t + 2$, and try to find the appropriate $\underset{\sim}{x}$, and hence valuate the weight of the codeword having $\underset{\sim}{Y}$ as its first m entries. This method will till include codewords of weight $\geq 2t + 2$, as the weight contributed from the second alf of the codeword must be added to the weight of $\underset{\sim}{Y}$, but at least some ordering has een achieved. So, we wish to solve for $\underset{\sim}{x}$, given $\underset{\sim}{Y}$:

$$(x_1,x_2,\ldots,x_{4t-1}) \begin{pmatrix} 0 & 1 & . & . & . & . & 1 \\ 1 & 0 & 1 & . & . & . & 1 \\ & & & . & & & \\ & & & . & & & \\ & & & . & & & \\ 1 & 1 & . & . & 1 & 0 & 1 \\ 1 & . & . & . & . & 1 & 0 \end{pmatrix} = (Y_1,Y_2,\ldots,Y_{4t-1}),$$

$$(x_1,x_2,\ldots,x_{4t-1}) \begin{pmatrix} 2 & 0 & . & . & . & 0 & 1 \\ 0 & 2 & 0 & . & . & 0 & 1 \\ & & & . & & & \\ & & & . & & & \\ 0 & . & . & . & 0 & 2 & 1 \\ 1 & . & . & . & 1 & 1 & 0 \end{pmatrix} = (Y_1 - Y_{4t-1},\ldots,Y_{4t-2} - Y_{4t-1},Y_{4t-1}),$$

$$(x_1,x_2,\ldots,x_{4t-1}) \begin{pmatrix} 2 & 0 & . & . & . & 0 & 0 \\ & & & . & & & \\ & & & . & & & \\ 0 & . & . & . & 0 & 2 & 0 \\ 1 & . & . & . & . & 1 & 0 \end{pmatrix} = (Y_1 - Y_{4t-1},\ldots,Y_{4t-2} - Y_{4t-1}, \sum_{i=1}^{4t-1} Y_i).$$

Notice that, because $(4t-1) - 1 \equiv 0 \pmod{3}$ for $t \equiv 2 \pmod{3}$, the $(4t-1, 4t-1)$ position of the last matrix is zero. Hence,

$$\text{rank } (G) \geq 4t - 2.$$

Also, since every column of A contains $(2t-1)$ 1's, and since $(2t-1) \equiv 0 \pmod{3}$ for $t \equiv 2 \pmod{3}$, the sum of all the rows of $A \equiv \underset{\sim}{0} \pmod{3}$.

$$\therefore \text{ rank } (G) = 4t - 2$$

$$\therefore \qquad k = 4t - 2.$$

Now, from the last matrix equation, a consistency check is placed on the possibilities of $\underset{\sim}{Y}$:

$$\sum_{i=1}^{4t-1} Y_i \equiv 0 \pmod{3}.$$

Also, as we need only test up to weight $2t + 1$, $\underset{\sim}{Y}$ will always contain at least one zero. Therefore, without loss of generality, by the circulant nature of A, we can set

$$Y_{4t-1} = 0.$$

This leaves:

$$(x_1, x_2, \ldots, x_{4t-1}) \begin{pmatrix} 2 & 0 & . & . & . & 0 & 0 \\ 0 & 2 & 0 & . & . & 0 & 0 \\ & & . & & & & \\ & & . & & & & \\ & & . & & & & \\ 0 & . & . & . & 0 & 2 & 0 \\ 1 & 1 & . & . & . & 1 & 0 \end{pmatrix} = (Y_1, Y_2, \ldots, Y_{4t-2}, \sum_{i=1}^{4t-2} Y_i).$$

solving this gives:

$$2x_i + x_{4t-1} = Y_i, \quad i = 1, 2, \ldots, 4t - 2. \qquad (*)$$

After solving this for x_i, we can then form the codeword $\underset{\sim}{x}[J-I \ A]$, and check to see if its weight is less than $2t + 2$. x_{4t-1} only appears in (*) because of the linearly dependent row in the matrix. It can be ignored, as it only contributes the solutions $\underset{\sim}{x} + 1$ and $\underset{\sim}{x} + 2$ to the solution $\underset{\sim}{x}$, which all yield the same codeword. This makes inversion trivial, as we can set x_{4t-1} equal to zero, and so we are left with:

$$2x_i = Y_i, \quad i = 1, 2, \ldots, 4t - 2. \qquad (**)$$

If $\underset{\sim}{\bar{x}}$ is the solution for $\underset{\sim}{x}$ in (**), then to obtain the vector $\underset{\sim}{Y}$, simply form:

$$\underset{\sim}{Y} = \underset{\sim}{\bar{x}}[J-I].$$

Thus, the method to find the minimum weight, is to find all possible vectors $\underset{\sim}{Y}$ (i.e. keeping the consistency check in mind), up to weight $2t + 1$, and to look through all the codewords:

$$\underset{\sim}{x}[J-I \ A]$$

resulting from the different $\underline{Y}$, for codewords of weight $\leq 2t + 1$. If none are found, there can be none, as every other codeword has the weight of its first m entries being $\geq 2t + 2$ and so clearly its total weight is $\geq 2t + 2$.

This method has the additional advantage, that if a codeword is found with weight $w < 2t + 2$, we then need only check up to weight $(w-1)$.

2. The second method makes use of the doubly circulant nature of G. To do this, notice that if we know:

$$\text{weight}(\underline{g}_i + \underline{g}_j + \underline{g}_k + \dots) \geq 2t + 2,$$

then we also know that:

$$\text{weight}(\underline{g}_{i+a} + \underline{g}_{j+a} + \underline{g}_{k+a} + \dots) \quad 2t + 2, \quad a = 1, \dots, 4t - 2,$$

and so these need not be searched through for a vector with weight $\leq 2t + 1$. The method for determining which rows do need to be checked, may be found in [2]. This is a very powerful method for obtaining the complete weight enumerator or the minimum distance, reducing the computation by a factor of m. It also applies equally well for the 2 cases $t \equiv 0$ and 1 (mod 3).

3. One further aid, which does not apply if $t \equiv 2$ (mod 3), is that iff A has full rank, then no linear combination of rows of A can be zero, and so all these linear combinations have weight ≥ 1 (e.g. it can be shown that for codes with generator matrix G, but $t \equiv 1$ (mod 3), A has full rank). In such a case, it would only be necessary to check up to weights of $\underline{Y}$ that are $\leq 2t$. Of course, in such a situation, G would have rank m, and so no consistency condition would exist.

RELATION TO [I H] CODE

We now show that G is contained within a code generated by a matrix of the form [I H].

Form the matrix

$$B = A - 2J.$$

Then

$$K = \begin{pmatrix} & 2 & \dots & 2 \\ I & \vdots & B & \\ & 2 & & \end{pmatrix}$$

is of the form [I H]. Delete the first row and column of K, the yield K_1. Add the last row of K_1 to two times each of the other rows in turn. Now, delete the last row of K_1, and its 4t th column, which contains only zeros, and call this K_2. The first 4t - 1 columns of K_2 are now identical to those we obtained by reducing G, where we can ignore the linearly dependent row. So, it remains to show that the rest of K_2 is also identical.

We only change B in forming K_2 from K_1. Consider $\underset{\sim}{a}_i$ and $\underset{\sim}{b}_i$, the ith rows of A and B respectively. We form $\underset{\sim}{k}_i$, the ith row of K_2 by:

$$\underset{\sim}{k}_i = 2\underset{\sim}{b}_i + \underset{\sim}{b}_{4t-1}$$
$$= 2\underset{\sim}{a}_i + \underset{\sim}{2} + \underset{\sim}{a}_{4t-1} + \underset{\sim}{1}$$
$$= 2\underset{\sim}{a}_i + \underset{\sim}{a}_{4t-1}$$

This is precisely what happens to the rows of A in the reduction of G.

SPECIFIC EXAMPLES

Applying the first two methods to the case where $t = 2$ and A has first row:

$$0 \quad 1 \quad 1 \quad 0 \quad 1 \quad 0 \;,$$

and to the case where $t = 5$ and A has first row

$$1 \quad 0 \quad 0 \quad 1 \quad 1 \quad 1 \quad 1 \quad 0 \quad 1 \quad 0 \quad 1 \quad 0 \quad 0 \quad 0 \quad 0 \quad 1 \quad 1 \quad 0 \quad 0,$$

we find that the upper bound of $2t + 2$ is attained in both cases. Hence we have a $(14,6,6)$ code and a $(38,18,12)$ code over $GF(3)$.

References

[1]   Wallis, W.D., Street, A.P., and Wallis, J.S., *Combinatorics: Room Squares, sum free sets and Hadamard Matrices*, in Lecture Notes in Mathematics, Vol. 292, Springer-Verlag, Berlin-Heidelberg-New York, 1972.

[2]   Razen, R., Seberry, J., and Wehrhahn, K., Ordered partitions and codes generated by circulant matrices, *J. Combinatorial Theory*, Ser. A.

[3]   Mallows, C.L., Plass, V., and Sloane, N.J.H., Self-dual codes over GF(3), *SIAM Journal of Applied Mathematics*, Vol. 31, No. 4, 1976.

[4]   MacWilliams, F.J., and Sloane, N.J.A., *The Theory of Error-Correcting Codes*, North Holland, Amsterdam-New York-Oxford, 1977.

Department of Applied Mathematics,
University of Sydney,
Sydney, 2006,
New South Wales,
Australia.

AN APPLICATION OF RENEWAL SEQUENCES TO THE DIMER PROBLEM

D.G. ROGERS

The theory of renewal sequences is used to obtain some exact results for the number of dimer and monomer-dimer configurations on blocks and rods in the hypercubical lattice and these results, in turn, are applied to give lower bounds for the dimer problem on the unrestricted lattice.

1. INTRODUCTION

Lattice statistics are of considerable interest in several areas in theoretical physics and much computational work has been done since the enumerative problems involved are often extremely difficult. The dimer problem is one of the most outstanding of these classical problems. A *dimer* is a pair of lattice points (sites) a unit distance apart in the d-dimensional hypercubical lattice and a *monomer* is a single lattice point. A *monomer-dimer (resp. dimer) configuration* on a (finite) subset S of the lattice is a partition of S into monomers and dimers (resp. dimers). The *dimer problem* is then to determine the number of such configurations on S when the dimers occupy a specified proportion of the lattice points of S. Detailed accounts of the problem are given in [16, 17] and further information and references are contained in the introductory sections of [5,7,8,9] (the first of these is the most recent and presents some fresh exact results).

For the most part, we are concerned with the case when no monomers are allowed (but see §6). For $\underline{a} = (a_1,\ldots,a_d)$, an $\underline{a}$-block is a d-dimensional parallelepiped, the edges of which in the i-th dimension contain a_i, $1 \le i \le d$, lattice points. The volume $N = N(\underline{a})$ of an $\underline{a}$-block is $N = a_1 a_2 \ldots a_d$. For N even, let $v(\underline{a})$ be the number of distinct dimer configurations on an $\underline{a}$-block. Then $v(\underline{a})$ is supermultiplicative in each of its arguments so that the limit

$$\lambda_d = \lim_{\underline{a} \to \infty} \log v(a)/N(\underline{a}) \qquad (1)$$

exists as a constant λ_d depending only on the dimension d of the lattice, $\underline{a} \to \infty$ indicating that separately for each i, $a_i \to \infty$ [6]. The problem is then to determine the constant λ_d.

In the case d = 2, we have the exact solution:-

$$\lambda_2 = \frac{1}{\pi} \sum_{r=0}^{\infty} \frac{(-1)^r}{(2r+1)^2} = 0.29156090\ldots . \qquad (2)$$

For d = 3, which is the case of greatest physical interest, we have the bounds

$$0.418347 \leq \lambda_3 \leq 0.548271 \qquad (3)$$

as special cases of general bounds for λ_d [7,15]. Further work (see for example [8]) has, however, been hampered by lack of knowledge of the evaluation of permanents and Pfaffians used in obtaining (2,3). So it is of some interest to find an alternative approach to the problem.

An elementary dissection argument is described in §3 which shows that $v(\underline{a})$ satisfies a family of renewal relations (for definitions see §2). By an application of the theory of these relations we may easily obtain bounds for λ_d, at least for small d. For example, for d = 2,3, we find:-

$$\lambda_2 \geq 0.2698 \;;\; \lambda_3 \geq 0.3505 \qquad (4)$$

The bounds (4) are weak in comparison with (2,3) but their derivation is much less technical and they may be improved by further computation. The renewal technique is also applied, in §5, to the determination of the number A(n,m) of monomer-dimer configurations on a (2,n)-block with m dimers (and so 2(n-m) monomers) which is another quantity of some physical interest [13].

The method employed here is akin to (and, in part, justifies) the work of Fowler and Rushbrooke [3] who, however, did not have access to the theory of renewal relations. There is also some similarity with the use of power series in tackling other physical problems [2], with the transfer matrix method [12], and with the work of Kesten [10] on self-avoiding walks. The method also has more general combinatorial applications [20].

We begin with a resumé of the theory of renewal relations (for a fuller account see [11, ch.1]).

2. RENEWAL RELATIONS

A sequence $\{u_n : n \geq 0\}$ is a *generalized renewal sequence* if

$$u_0 = 1 \;;\; 0 \leq u_n, \quad n \geq 1 \qquad (5)$$

and, for some non-negative sequence $\{f_n : n \geq 1\}$, it satisfies the *renewal relation*:-

$$u_n = \sum_{r=1}^{n} f_r u_{n-r}, \quad n \geq 1 \qquad (6)$$

or, on introducing generating functions, the formal identity:-

$$U(x) = \sum_{n \geq 0} u_n x^n = 1 + U(x)F(x) \;;\; F(x) = \sum_{n \geq 1} f_n x^n.$$

It is clear, from (6), that the sequences $\{u_n : n \geq 0\}$ and $\{f_n : n \geq 1\}$ may be determined from one another. The *period* p of a generalized renewal sequence $\{u_n : n \geq 0\}$ is given by

$$p = \text{g.c.d.} \{n \geq 1 : u_n > 0\}$$

here, for a set T of integers, g.c.d. T stands for the greatest common divisor of the
embers of T. For our present purposes, it is sufficient to consider only the *aperiodic*
ase, p = 1.

If a sequence $\{u_n : n \geq 0\}$ in addition to (5,6) also satisfies

$$u_n \leq 1 \tag{7}$$

r, as may be shown, equivalently

$$\sum_{n\geq 1} f_n \leq 1 \tag{8}$$

hen the sequence is said to be a *renewal sequence*. It is, in fact, renewal sequences
hich are the more usual objects of study since they arise in the theory of Markov
hains [11, p.5], conditions (7) or (8) being essentially probabilistic. We have intro-
uced generalized renewal sequences and renewal relations here as a slight generalization
o cover the present application and other combinatorial situations where (7) and (8)
re less appropriate [20,21] (see also [23]). We now extend the well known limit results
1,11] for renewal sequences to generalized renewal sequences.

Thus, if (5,6) are satisfied, then the non-negativity of the f_n implies that the
$_n$ are supermultiplicative

$$u_m u_n \leq u_{m+n}, \quad n,m \geq 0$$

s may be seen by multiplying (6) by u_m and using induction on m + n. In the aperiodic
ase, this in turn implies [11, p.7] that the limit

$$u = \lim_{n\to\infty}(u_n)^{1/n} \tag{9}$$

xists with $0 \leq u \leq +\infty$ and

$$e^{n\log u + 0(n)} = u_n \leq u^n, \quad n \to \infty. \tag{10}$$

ence, if the limit is finite, as it will be if, for example, there are constants a,b,
nd k such that

$$u_n \leq ak^n + b, \quad n \geq 1$$

hen, in view of (10), the sequence $\{v_n : n \geq 0\}$ given by

$$v_n = u_n/u^n, \quad n \geq 0$$

s a renewal sequence. The limit results [1] for renewal sequences now carry over to
how that

$$1 = \lim_{n\to\infty}(v_n)^{1/n} = \text{l.u.b.} \{x \geq 0 : F(x/u) \leq 1\}$$

$$F(1/u) \leq 1 \tag{11}$$

where equality holds in (11) if $U(1/u)$ diverges. Moreover [11, p.12]

$$\lim_{n\to\infty} (u_n/u^n) = \lim_{n\to\infty} v_n = u/F'(1/u) \tag{12}$$

with the interpretation that the right hand side of (12) is zero if $F'(1/u)$ diverges.

If the sequence $\{u_n : n \geq 0\}$ satisfies a linear difference equation with constant coefficients, as happens in some special cases in §4, then equality holds in (11). If equality does hold in (11) then $1/u$ is characterized as the unique positive root of the equation

$$F(x) = 1.$$

Even if equality does not hold in (11), it is still possible to obtain bounds for u from (6). Let k be a positive integer and define a sequence $\{u_n^{(k)} : n \geq 0\}$ by

$$u_n^{(k)} = u_n \ , \ n < k \ ; \ = \sum_{r=1}^{k} f_r u_{n-r}^{(k)} \ , \ n \geq k$$

so, by induction, $u_n^{(k)} \leq u_n \leq u^n$, $n \geq 1$, and, in addition, by the previous remarks,

$$\lim_{n\to\infty} (u_n^{(k)})^{1/n} = u^{(k)} \ , \ \text{say},$$

where $u^{(k)}$ is the largest positive root of

$$\sum_{r=1}^{k} f_r/x^r = 1.$$

The sequence $\{u^{(k)} : k \geq 1\}$ is monotonic increasing with limit u so, by Fatou's lemma we obtain (11) and the $u^{(k)}$ provide successively better bounds for u.

3. A DISSECTION ARGUMENT

Regarding $v(\underset{\sim}{a})$ as a function of, say, its last argument, we may write

$$u_n = v_n(a_1,\ldots,a_{d-1}) = v(\underset{\sim}{a}_n), \quad n \geq 1 \tag{13}$$

where $\underset{\sim}{a}_n = (a_1,\ldots,a_{d-1},n)$. We may consider the $\underset{\sim}{a}_n$-block as being made up of n parallel a_1 by a_2 by ... by a_{d-1} (d - 1) dimensional slices. Let $f_n = f_n(a_1,\ldots,a_{d-1})$ be the number of dimer configurations on the block such that each slice is linked to its neighbouring slices by at least one dimer. If two neighbouring slices are not linked, then by 'cutting' between them we may divide the block into two sub-blocks on each of which we have a dimer configuration. Since, for any configuration, there is a unique slice nearest the right hand end at which the configuration may be so split and there are $f_r u_{n-r}$ dimer configurations on the $\underset{\sim}{a}_n$-block for which the first split occurs immediately after the r-th slice, $1 \leq r \leq n$, we have, setting $u_0 = 1$, (compare 6)

$$u_n = \sum_{r=1}^{n} f_r u_{n-r}, \quad n \geq 1. \tag{14}$$

onversely starting from u_n and subtracting the number of configurations which break
efore the n-th slice we obtain f_n. Since the f_n are non-negative, this shows that
$u_n : n \geq 0\}$ is a generalized renewal sequence.

A completely recursive method of calculating both u_n and f_n may also be given based
n the method of dissection used in the previous paragraph. Let $\underset{\sim}{a}_n$ be as above with,
or convenience, $N^* = \frac{1}{n} N(\underset{\sim}{a}_n)$ even. Then an $\underset{\sim}{a}_{m+n}$-block may be regarded in a natural way
s the union of an $\underset{\sim}{a}_m$-block A, say, with an $\underset{\sim}{a}_n$-block B. First join A to B at their ad-
acent faces by k dimers and then, if possible, fill the remainder of A and B with dimers
since N^* is even, k must also be even). For $n,m \geq 1$, and u_n as in (13), if $k = 0$, the
umber of ways of doing this is just $u_m u_n$ while if $k = N^*$, it is $u_{m-1} \cdot u_{n-1}$; otherwise
t is

$$\sum_{\omega \in \Omega_k} v_m^\omega v_n^\omega$$

here Ω_k indexes the ways ω in which k out of N^* lattice points comprising the face of
adjacent to B may be removed and v_m^ω is the number of ways of partitioning A into dimers
nce k points have been removed from this face in way ω. Hence, since all configurations
n the original block produced in this way are distinct and the decomposition is exhaust-
ve, it follows that

$$u_{m+n} = u_m u_n + u_{m-1} u_{n-1} + \sum_{\omega \in \Omega} v_m^\omega v_n^\omega, \quad m,n \geq 1,$$

here $\Omega = \bigcup_{k=2}^{N^*-2} \Omega_k$. In particular, taking $m = 1$ and noting that $u_0 = 1$, we have (compare
6,14))

$$u_{n+1} = u_1 u_n + u_{n-1} + \sum_{\omega \in \Omega} b^\omega v_n^\omega \tag{15}$$

or some constants $b^\omega = v_1^\omega$.

In exactly the same way, the v_n^ω satisfy linear equations

$$v_n^\omega = c_1(\omega) + \sum_{\omega' \in \Omega} c(\omega,\omega') v_{n-1}^{\omega'} \tag{16}$$

or some constants $c_1(\omega)$ and $c(\omega,\omega')$ so that, on iteration, we have

$$v_n^\omega = \sum_{i=1}^{n} c_i(\omega) u_{n-i}$$

or some further constants $c_i(\omega)$ determined recursively by a family of linear equations
imilar to and derived from (16). Substituting these expressions into (15) and compar-
ng with (14), we find

$$f_1 = u_1; \ f_2 = 1 + \sum_{\omega \in \Omega} b(\omega) c_1(\omega)$$

$$f_{n+1} = \sum_{\omega \in \Omega} b(\omega) c_n(\omega), \quad n \geq 2. \tag{17}$$

Finally, since by superposition,

$$u_n \leq u_2^{n-1},$$

the limit in (9) is finite and the limit results of the previous section may be applied.

4. SOME EXACT RESULTS

For small d and small fixed a_i, $i < d$, when there is little room for manoeuvre, we may determine the sequence $\{f_n : n \geq 1\}$ explicitly. We illustrate the method of the preceding section by considering the case of a 4 by (n + 1) planar block P of lattice points, that is $d = 2$, $a_1 = 4$, $a_2 = n + 1$. Label the lattice points in one of the edges of P consisting of four lattice points consecutively by the numbers 1,2,3,4. Let a_{n+1}, b_{n+1}, and c_{n+1} be the number of dimer configurations on $P\backslash\{1,2\}$, $P\backslash\{1,4\}$, and $P\backslash\{2,3\}$, respectively. Then it follows, after the manner of (15,16) and using the symmetry of P, that

$$u_{n+1} = v(4, n + 1) = u_n + u_{n-1} + a_n + 2b_n, \quad n \geq 1$$

where

$$a_{n+1} = u_n + c_n \; ; \; b_{n+1} = u_n + b_n \; ; \; c_{n+1} = a_n, \quad n \geq 1.$$

So, comparing with (17), since $a_1 = b_1 = 1$ and $c_1 = 0$, the sequence $\{f_n : n \geq 1\}$ is given explicitly by $f_1 = 1$, $f_2 = 4$ and

$$f_{2n-1} = 2 \; , \; f_{2n} = 3, \quad n \geq 2. \tag{18}$$

The configurations on (4,3)- and (4,4)-blocks having no (4,r)-sub-block configurations are illustrated in Figures (1a) and (1b) respectively, showing that $f_3 = f_3(4) = 2$ and $f_4 = f_4(4) = 3$ in agreement with (18).

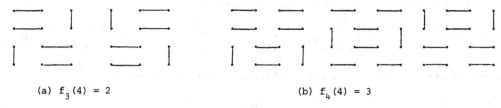

(a) $f_3(4) = 2$ (b) $f_4(4) = 3$

Figure 1

Other easily determined explicit examples are:-

d = 2 ; $a_1 = 2$: $f_1 = f_2 = 1$; $f_n = 0$, $n \geq 3$ (19)

d = 3 ; $a_1 = a_2 = 2$: $f_1 = 2$; $f_2 = 5$; $f_n = 4$, $n \geq 3$.   (20)

From (18,19,20), we obtain the recurrence relations:-

$$= 2 \; ; \; a_1 = 4 : \qquad u_n - u_{n-1} - 5u_{n-2} - u_{n-3} - u_{n-4} = 0 \; , \qquad n \geq 4$$

$$= 2 \; ; \; a_1 = 2 : \qquad u_n - u_{n-1} - u_{n-2} = 0 \; , \qquad\qquad\qquad n \geq 2 \qquad (21)$$

$$= 3 \; ; \; a_1 = a_2 = 2 : \qquad u_n - 3u_{n-1} - 3u_{n-2} + u_{n-3} = 0 \; , \qquad n \geq 3.$$

n each of these cases, equality holds in (11) and the limit result (12) also holds.
he limit u of (9) may be identified as the largest positive root of the auxiliary
quation associated with the recurrence relation and numerical calculations give the
ollowing values in these cases:-

$$d = 2 \; ; \; a_1 = 4 : u = 2.8405... \qquad (22)$$

$$d = 3 \; ; \; a_1 = a_2 = 2 : u = 3.7320... \qquad (23)$$

The recursive method of the previous section does however, even after taking account
f the symmetries of the $\underset{\sim}{a}$-blocks, become somewhat unwieldly as it involves large numbers
f linear equations (16). For example, for $d = 2$, there are twelve equations when $a_1 = 6$
nd forty when $a_1 = 8$ while, for $d = 3$, $a_1 = 2$, there are eight when $a_2 = 3$ and twenty
hen $a_2 = 4$. Some calculations were made in these cases and, using the method indicated
t the end of the second section, the following results were obtained:-

$$d = 2 \; ; \; a_2 = 6 : u > 5.0472 \qquad (24)$$

$$d = 3, \; a_1 = 2 \; ; \; a_2 = 3 : u > 7.7954$$

$$d = 3, \; a_1 = 2 \; ; \; a_2 = 4 : u > 16.5100. \qquad (25)$$

. BOUNDS FOR THE DIMER PROBLEM

The limit $\frac{1}{N^*} \log u$, where u is as in (9) and $\{u_n : n \geq 0\}$ is the generalized renewal
equence obtained in §3, is the analogue for 'rods' in the lattice of the limit λ_d of
1) for the unrestricted lattice. Since the unrestricted lattice may be partitioned
nto rods, the limits obtained by the method of the previous two sections may be used
o give bounds for the λ_d. More precisely, the number of dimer configurations on an
$\underset{\sim}{a}$-block is greater than the number on n^{d-1} parallel $\underset{\sim}{a_n}$-blocks where $\underset{\sim}{a_n} = (a_1, \ldots, a_{d-1}, n)$
nd $\underset{\sim}{na_1} = (na_1, \ldots, na_{d-1}, n)$.
o

$$v(\underset{\sim}{na_1}) \geq (v(\underset{\sim}{a_n}))^{n^{d-1}} = (u_n)^{n^{d-1}}$$

rom which, since $N(\underset{\sim}{na_1}) = n^{d-1} N(\underset{\sim}{a_n}) = n^d N^*$, it follows that

$$\frac{1}{N(\underset{\sim}{na_1})} \log v(\underset{\sim}{na_1}) \geq \frac{1}{N^*} \left(\frac{1}{n} \log u_n \right).$$

ow, letting n tend to infinity and noting (1) and (9), we have

$$\lambda_d \geq \frac{1}{N^*} \log u.$$

Taking $\underset{\sim}{a}_1 = (4,1)$, we have from (22),

$$\lambda_2 \geq \frac{1}{4} \log u = 0.2609\ldots$$

while with $\underset{\sim}{a}_1 = (2,2,1)$, we obtain from (23),

$$\lambda_3 \geq \frac{1}{4} \log u = 0.3292\ldots$$

These bounds may be improved by using (24) and (25) to give (4).

Upper bounds for λ_d may be obtained by taking overlapping rods in the decomposition of the unrestricted lattice but these bounds seem to be very weak. We may also obtain, by similar arguments, limit results for the sequence $\{f_n : n \geq 1\}$. For example, for $d = 2$, we have

$$\lim_{n,m \to \infty} \frac{1}{nm} \log f_n(m) = \lambda_2$$

since

$$v(m - 4,n) \leq f_n(4) \, v(m - 4,n) \leq f_n(m) \leq v(m,n), \quad m \geq 4.$$

Further, if $g(n,m)$ is the number of dimer configurations on an m by n planar block which cannot be 'cut' in either axial direction into configurations on sub-blocks without breaking a dimer, then bordering configurations on an m - 8 by n - 8 block with suitable 'unbreakable' configurations on strips gives

$$v(m - 8, n - 8) \leq (f_{n-4}(4) f_{m-4}(4))^2 v(m - 8, n - 8) \leq g(m,n) \leq v(m,n)$$

and so

$$\lim_{n,m \to \infty} \frac{1}{nm} \log g(n,m) = \lambda_2.$$

6. AN EXACT RESULT FOR (2,n)-BLOCKS

Interest has been shown recently [4,13,14,18] in determining, for n > 0, the number $A(n,m)$, $0 \leq m \leq n$, of monomer-dimer configurations on a (2,n)-block with m dimers (we also set $A(0,0) = 1$). When m = n the block is said to be *full* and, from §4, we have for $n \geq 1$,

$$A(n,n) = v(2,n) = F_n$$

where F_n is the n-th *Fibonacci number* [19] given by $F_0 = F_1 = 1$ and (compare (21))

$$F_n = F_{n-1} + F_{n-2}, \quad n \geq 2. \tag{26}$$

Various exact formulae for $A(n,m)$ have been obtained and the first few values are set out in the following table:

n\m	0	1	2	3	4	5
0	1					
1	1	1				
2	1	4	2			
3	1	7	11	3		
4	1	10	29	26	5	
5	1	13	56	94	56	8

The problem of determining $A(n,m)$ is also amenable to the renewal approach and this approach reveals a multiplicative structure in the array $\{A(n,m) : 0 \le m \le n\}$ not hitherto noticed: in the terminology of [21,22], the array is a forced renewal array.

The *principal left (right) (2,r)-block*, $1 \le r \le n$, of a (2,n)-block is the (2,r)-block at the left (right) hand end of the (2,n)-block. A *critical configuration* on a (2,n)-block is a monomer-dimer configuration with $n - 1$ dimers (and so two monomers) in which none of the principal left (2,r)-blocks, $1 \le r \le n$, is full. To exhibit the multiplicative structure alluded to above, we let B_n, $n \ge 1$, be the number of critical configurations on a (2,n)-block. Then, firstly, B_n satisfies a recurrence relation similar to (26). For, given a critical configuration on a (2,n)-block, $n \ge 3$, a split occurs either after the principal left (2,n-1)-block or after the principal left (2,n-2)-block or no split occurs and the dimers are arranged in one of two staggered configurations. Since these possibilities are exclusive and exhaustive, we have, for $n \ge 3$,

$$B_n = B_{n-1} + B_{n-2} + 2$$

so, since $B_1 = 1$ and $B_2 = 3$, the first few values are:-

n	1	2	3	4	5	6	7	8
B_n	1	3	6	11	19	32	53	87

The B_n are related to the Fibonacci numbers F_n by

$$B_n = \sum_{r=1}^{n} F_r, \qquad n \ge 1,$$

and we have the explicit expression

$$B_n = \frac{1}{\sqrt{5}}\left((3 - 2\beta)\alpha^n - (3 - 2\alpha)\beta^n \right) - 2, \qquad n \ge 1,$$

where

$$\alpha = \frac{1 + \sqrt{5}}{2}; \quad \beta = \frac{1 - \sqrt{5}}{2}.$$

The sequence $\{B_n : n \ge 1\}$ also occurs in [19, p.233].

Next, consider a monomer-dimer configuration on a (2,n)-block with m dimers, $m < n$. Then there is a greatest integer i, $1 \le i \le m + 1$, such that the configuration splits into a critical configuration on the principal right (2,i)-block and a configuration on the principal left (2,n-i)-block with $m - i + 1$ dimers. Hence, taking account of all these possibilities,

$$A(n,m) = \sum_{i=1}^{m+1} A(n - i, m - i + 1)B_i, \qquad 0 \le m \le n. \qquad (27)$$

Now, introducing the generating functions

$$A^{(r)}(x) = \sum_{n \ge 0} A(n + r - 1, n)x^n, \qquad r \ge 1$$

$$A(x) = A^{(1)}(x) \; ; \; B(x) = \sum_{n \ge 1} B_n x^{n-1},$$

we have, from (27)

$$A^{(r)}(x) = A^{(r-1)}(x) \, B(x), \quad r \geq 2.$$

Hence, inductively,

$$A^{(r)}(x) = A(x)\left(B(x)\right)^{r-1}, \quad r \geq 1,$$

which demonstates the multiplicative structure of the array $\{A_{(n,m)} : 0 \leq m \leq n\}$ mentioned earlier.

REFERENCES

[1] N.G. de Bruijn and P. Erdös, 'Some linear and some quadratic recursion formulas', *Indagationes Math.* 13 (1951), 374-382.

[2] C. Domb and M.F. Sykes, 'Use of series expansions for the Ising model susceptibilit and excluded volume problem', *J. Math. Phys.* 2 (1961), 63-67.

[3] R.M. Fowler and G.S. Rushbrook, 'Statistical theory of perfect solutions', *Trans. Faraday Soc.* 33 (1937), 1272-1294.

[4] R.C. Grimson, 'Exact formulas for 2 × n arrays of dumbbells', *J. Math. Phys.* 15 (1974), 214-216.

[5] R.C. Grimson, 'Enumeration of dimer (domino) configurations', *Discrete Math.* 18 (1977), 167-177.

[6] J.M. Hammersley, 'Existence theorems and Monte Carlo methods for the monomer dimer problem', in *Research papers in statistics: Festschrift für J. Neyman,* ed. F.N. David, J. Wiley, New York, 1966.

[7] J.M. Hammersley, 'An improved lower bound for the multidimensional dimer problem', *Proc. Camb. Phil. Soc.* 64 (1968), 455-463.

[8] J.M. Hammersley and V.V. Menon, 'A lower bound for the monomer-dimer problem', *J. Inst. Math. Applics.* 6 (1970), 341-364.

[9] O.J. Heilmann and E.H. Lieb, 'Theory of monomer-dimer systems', *Comm. Math. Phys.* 25 (1972), 190-232.

[10] H. Kesten, 'On the number of self avoiding walks', *J. Math. Phys.* 4 (1963), 960-969

[11] J.F.C. Kingman, *Regenerative Phenomena,* J. Wiley, London, 1972.

[12] E.H. Lieb, 'Solution of the dimer problem by the transfer matrix method', *J. Math. Phys.* 8(1967), 2339-2341.

[13] R.B. McQuistan and S.J. Lichtman, 'Exact recursion relation for 2 × n arrays of dumbbells', *J. Math. Phys.* (1970), 3095-3097.

[14] R.B. McQuistan, S.J. Lichtman, and L.P. Levine, 'Occupation statistics for parallel dumbbells on a 2 × N lattice space', *J. Math. Phys.* 13 (1972), 242-248.

[15] H. Minc, 'An upper bound for the multidimensional dimer problem', *Math. Proc. Camb. Phil. Soc.* 83 (1978), 461-462.

[16] E.W. Montroll, 'Lattice statistics', in *Applied Combinatorial Mathematics,* ed. F. Beckenback, J. Wiley, New York, 1964.

[17] J.K. Percus, *Combinatorial Methods*, Springer-Verlag, Berlin, 1971.

[18] M.A. Rashid, 'Occupation statistics from exact recursion relations for occupation by dumbbells of a 2 × n array', *J. Math. Phys.* 15 (1974), 474-476.

[19] J. Riordan, *Introduction to Combinatorial Analysis*, J. Wiley, New York, 1958.

[20] D.G. Rogers, 'Pascal triangles, Catalan numbers, and renewal arrays', *Discrete Math.* 22(1978), 301-311.

[21] D.G. Rogers, 'Lattice paths with diagonal steps.'

[22] D.G. Rogers, 'Similarity relations and semiorders.'

[23] D.N. Shanbhag, 'On renewal sequences', *Bull. London Math. Soc.* 9 (1977), 79-80.

Department of Mathematics,
University of Western Australia,
Nedlands, 6009
Western Australia,
Australia.

SOME REMARKS ON GENERALISED HADAMARD MATRICES AND THEOREMS OF RAJKUNDLIA ON SBIBDS

JENNIFER SEBERRY

Constructions are given for generalised Hadamard matrices and weighing matrices with entries from abelian groups.

These are then used to construct families of SBIBDs giving alternate proofs to those of Rajkundlia.

1. DEFINITION

A *generalised Hadamard matrix* GH(n,G) is an n × n matrix with elements from the abelian group G of order $|G|$ such that if $\underset{\sim}{a} = (a_1,\ldots,a_n)$ and $\underset{\sim}{b} = (b_1,\ldots,b_n)$ are any two rows of GH(n,G) then the elements $a_i b_i^{-1}$, i = 1,...,n give n/$|G|$ copies of G. These matrices were considered by Butson [4,5], by Shrikhande [18] in connection with combinatorial designs, by Delsarte and Goethals [6,7] in connection with codes and Drake [8] in connection with λ- geometries.

A *generalised weighing matrix* GW(n,k,G) is an n × n matrix with elements from the abelian group G of order $|G|$ and zero, there are k non-zero elements per row and column and if $\underset{\sim}{a} = (a_1,\ldots,a_n)$ and $\underset{\sim}{b} = (b_1,\ldots,b_n)$ are any two rows of GW(n,k,G) then the elements $a_i b_i^{-1}$, i = 1,...,n give λ_{ab} copies of G. If λ_{ab} is a constant for all a and b we have a *balanced weighing matrix*.

Weighing matrices, the special case with G the cyclic group of order 2 have been studied extensively [10,11,13,19,22]. Their name comes from Yates [25] who gave an application in the accuracy of measurements. Balanced weighing matrices have been studied in connection with combinatorial designs by Mullin and Stanton [14,15,16,21] and Berman [2]. Complex weighing matrices have been studied by Berman [3] and Geramita and Geramita [9].

To illustrate that Berman's generalised weighing matrices and ours are not the same we consider

$$A = \begin{bmatrix} 0 & 1 & 1 & i \\ 1 & 0 & i & 1 \\ i^2 & i & 0 & 1 \\ i & i^2 & 1 & 0 \end{bmatrix}$$

which satisfies AA* = 3I and is a W(4,3,Z_4) when $i^2 = -1$ but is not a generalised weighing matrix by our definition as the product of rows 1 and 2 is {i,i^3} and we need {$1,i,i^2,i^3$}.

Notation. Throughout this paper we use Z_q for the cyclic group on q symbols and C_{p^r} for the elementary abelian group $Z_p \times Z_p \times \ldots \times Z_p$.

For our purposes an SBIBD(v,k,λ) is a matrix with entries 0 and 1 of order v with k ones per row and column and inner product between rows of λ.

David Glynn [12] has found the only GW(v,k,G) known to the author where G is not an abelian group. Consider the multiplication table for S_3

	1	2	3	4	5	6		
1	1	2	3	4	5	6	$1 \leftrightarrow e$	
2	2	3	1	5	6	4	$2 \leftrightarrow (123)(456)$	
3	3	1	2	6	4	5	$3 \leftrightarrow (132)(465)$	
4	4	6	5	1	3	2	$4 \leftrightarrow (14)(26)(35)$	
5	5	4	6	2	1	3	$5 \leftrightarrow (15)(24)(36)$	
6	6	5	4	3	2	1	$6 \leftrightarrow (16)(25)(34)$	

Then the circulant matrix with first row

$$[0\ 5\ 1\ 4\ 0\ 1\ 1\ 6\ 5\ 6\ 0\ 4\ 0]$$

is a generalised weighing matrix $GW(13,9,S_3)$.

2. A FAMILY OF GENERALISED WEIGHING MATRICES

We first give a more direct construction for a result implicit in the work of Rajkundlia. We note that our matrix implies the one of Berman but has an additional property and is obtained quite differently.

Let γ be a primitive element of $GF(p^r)$. Let $q \mid p^r - 1$ and let α be a generator of Z_q, the cyclic group. Write $g_1 = 0, g_2,\ldots,g_{p^r}$ for the elements of $GF(p^r)$ and define $M = (m_{ij})$ of order $p^r + 1$ as follows:

$$m_{ii} = 0$$
$$m_{ij} = \alpha^k \qquad \text{where } g_j - g_i = \gamma^k$$
$$m_{0j} = m_{j0} = 1.$$

Example. Let γ be a primitive element of $GF(2^2)$ and ω be a primitive element of $GF(3)$ where $q = 3$. Write $g_1 = 0$, $g_2 = 1$, $g_3 = \gamma$, $g_4 = \gamma + 1$ for the elements of $GF(2^2)$ using $\gamma^2 = \gamma + 1$. Now

$$M = \begin{bmatrix} 0 & 1 & 1 & 1 & 1 \\ 1 & 0 & 1 & \omega & \omega^2 \\ 1 & 1 & 0 & \omega^2 & \omega \\ 1 & \omega & \omega^2 & 0 & 1 \\ 1 & \omega^2 & \omega & 1 & 0 \end{bmatrix}$$

We note that M is a $GW(5,4,Z_3)$.

Example. Let $\gamma = 3$ be a primitive element of $GF(7)$ and ω be a primitive element of $GF(3)$ where $q = 3$. Write $g_i = i - 1$, $i = 1,\ldots,7$ for the elements of $GF(7)$. Now

$$M = \begin{bmatrix} 0 & 1 & 1 & 1 & 1 & 1 & 1 & 1 \\ 1 & 0 & 1 & \omega^2 & \omega & \omega & \omega^2 & 1 \\ 1 & 1 & 0 & 1 & \omega^2 & \omega & \omega & \omega^2 \\ 1 & \omega^2 & 1 & 0 & 1 & \omega^2 & \omega & \omega \\ 1 & \omega & \omega^2 & 1 & 0 & 1 & \omega^2 & \omega \\ 1 & \omega & \omega & \omega^2 & 1 & 0 & 1 & \omega^2 \\ 1 & \omega^2 & \omega & \omega & \omega^2 & 1 & 0 & 1 \\ 1 & 1 & \omega^2 & \omega & \omega & \omega^2 & 1 & 0 \end{bmatrix}$$

We note M is a $GW(8,7,Z_3)$.

Theorem 1. *Suppose p^r is a prime power and $q \mid p^r - 1$. Then there exists a balanced $GW(p^r+1, p^r, Z_q)$.*

Proof. Construct M of order $p^r + 1$ as above. We show M is the required $GW(p^r+1, p^r, Z_q)$. First M has the elements $0, 1$ $((p^r-1)/q + 1$ times), and $\alpha, \alpha^2, \ldots, \alpha^{q-1}$ (each $(p^r-1)q$ times) in each row (column) but the first. So we have the group property with respect to the first row.

We now consider the other rows. We consider $q = p^r - 1$. Suppose $g_j - g_i = \gamma^b$, $g_j - g_k = \gamma^s$ then $m_{ij} = \alpha^b$, $m_{kj} = \alpha^s$. We wish to show that $m_{ij} m_{kj}^{-1} = \alpha^{b-s}$ cannot arise in any other way. We proceed by reductio ad absurdum. Suppose there exists other entries so $g_m - g_i = \alpha^a$ and $g_m - g_k = \gamma^r$, where $m_{mi} = \gamma^a$ and $m_{mk} = \alpha^r$. That is, $m_{mi} m_{mk}^{-1} = \alpha^{a-r}$ where $a - r = b - s$. Then $g_k - g_i = \gamma^a - \gamma^r = \gamma^r(\gamma^{a-r}-1)$ and $g_k - g_i = \gamma^b - \gamma^s = \gamma^s(\gamma^{b-s}-1)$. So $s = r$ and $a = b$. But this means there were no other entries. Hence each of the $p^r - 2$ elements $m_{ij} m_{kj}^{-1}$, $i \neq j$, $k \neq j$, $j = 1, \ldots, q$ is different. It is not possible for $m_{ij} = m_{kj}$ so the $p^r - 2$ elements are $\alpha, \ldots, \alpha^{q-1}$. The 1 comes from $m_{i0} m_{k0}^{-1}$.

We saw that when $q = p^r - 1$ the $p^r - 2$ elements $m_{ij} m_{kj}^{-1}$, $i \neq j$, $k \neq j$, $j = 1, \ldots, q$ where $\alpha, \ldots, \alpha^{q-1}$. Hence if $q_1 \mid p^r - 1$ so $\alpha^{q_1} = 1$ these $p^r - 2$ elements will be $\alpha, \ldots, \alpha^{q_1-1}$ $((p^r-1)/q_1$ times) and 1 $((p^r-1)/q_1 - 1$ times). The additional 1 comes from $m_{i0} m_{k0}^{-1}$.

So we have a generalised $GW(p^r+1, p^r, Z_q)$. The matrix is balanced as the underlying SBIBD is (p^r+1, p^r, p^r-1).

Remark. This construction was first given for $q = 4$ in [19, p.297].

3. SOME GENERALISED HADAMARD MATRICES $GH(p^r, C_{pr})$ and $GH(p^r(p^r-1), C_{pr})$

The $GH(p^r, Z_p \times \ldots \times Z_p)$ was first noted by Drake [8] but we give it here for illustrative purposes.

Let x be a primitive element of $GF(p^r)$. We form

$$X = \left(x^{j-i+1 \pmod{p^r-1}} \right).$$

Now the generalised Hadamard matrix on the elementary abelian group in additive form is formed by reducing the elements of X modulo a primitive polynomial and adding a zeroth row and column which is the additive identity. This matrix can now be written multiplicatively to obtain $GH(p^r, Z_p \times Z_p \times \ldots \times Z_p)$.

For example, let x be a primitive element of $GF(3^2)$. We form

$$X = x \quad x^2 \quad x^3 \quad \ldots \quad x^8$$
$$x^8 \quad x \quad x \quad \ldots \quad x^7$$
$$\vdots$$
$$x^2 \quad x^3 \quad x^4 \quad \ldots \quad x$$

then the generalised Hadamard matrix using $x^2 = x + 1$ is

0	0	0	0	0	0	0	0	0
0	x	x+1	2x+1	2	2x	2x+2	x+2	1
0	1	x	x+1	2x+1	2	2x	2x+2	x+2
0	x+2	1	x	x+1	2x+1	2	2x	2x+2
$\vdots$								
0	x+1	2x+1	2	2x	2x+2	x+2	1	x

or in multiplicative form

1	1	1	1	1	1	1	1	1
1	a	ab	a^2b	b^2	a^2	a^2b^2	ab^2	b
1	b	a	ab	a^2b	b^2	a^2	a^2b^2	ab^2
1	ab^2	b	a	ab	a^2b	b^2	a^2	a^2b^2
$\vdots$								
1	ab	a^2b	b^2	a^2	a^2b^2	ab^2	b	a

The corresponding matrices, if x (=3) is a primitive element of $GF(5)$, are

x	x^2	x^3	x^4		0	0	0	0	0		1	1	1	1	1
x^4	x	x^2	x^3		0	3	4	2	1		1	a^3	a^4	a^2	a
x^3	x^4	x	x^2		0	1	3	4	2		1	a	a^3	a^4	a^2
x^2	x^3	x^4	x		0	2	1	3	4		1	a^2	a	a^3	a^4
					0	4	2	1	3		1	a^4	a^2	a	a^3

For reference purposes we note the following theorem. A direct proof of (ii), inspired by Rajkundlia, will appear elsewhere.

Theorem 2. *(i) Suppose p^r is a prime power. Then there is a $GH(p^r, C_{p^r})$* where C_{p^r} *is the elementary abelian group.*

      (ii) Suppose p^r and $p^r - 1$ are both prime powers. Then there is a $GH(p^r(p^r-1), C_{p^r})$ *where C_{p^r} is the elementary abelian group.*

Example of construction of $GH(12, Z_2 \times Z_2)$

	e	a	b	ab
e	e	a	b	ab
a	a	e	ab	b
b	b	ab	e	a
ab	ab	b	a	e

has core C =

e	ab	b
ab	e	a
b	a	e

The generalised Hadamard matrix of order 4:

$$
\begin{array}{cccc}
e & e & e & e \\
e & a & b & ab \\
e & b & ab & a \\
e & ab & a & b
\end{array}
\qquad
\text{has core } K =
\begin{array}{ccc}
a & b & ab \\
b & ab & a \\
ab & a & b
\end{array}
$$

Let I, T, T^2 of order 3 be a matrix representation of ε, w, w^2 where w is a cube root of unity, then

$$
W =
\begin{array}{ccc}
\varepsilon & \varepsilon & \varepsilon \\
\varepsilon & w & w^2 \\
\varepsilon & w^2 & w
\end{array}
$$

is a generalised Hadamard matrix of order 3.

Now define

$$
C*W =
\begin{array}{ccc}
e\varepsilon & ab\varepsilon & b\varepsilon \\
ab\varepsilon & ew & aw^2 \\
b\varepsilon & aw^2 & ew
\end{array}
$$

and

$$
D =
\begin{array}{ccc}
eK & abK & bK \\
abK & eKT & aKT^2 \\
bK & aKT^2 & eKT
\end{array}
$$

and the following is the required matrix:

$$
\begin{array}{cccccc}
e & e & e & \underset{\sim}{e} & \underset{\sim}{e} & \underset{\sim}{e} \\
e & e & e & \underset{\sim}{b} & \underset{\sim}{ab} & \underset{\sim}{a} \\
e & e & e & \underset{\sim}{ab} & \underset{\sim}{a} & \underset{\sim}{b} \\
\underset{\sim}{e}' & \underset{\sim}{b}' & \underset{\sim}{ab}' & eK & abK & bK \\
\underset{\sim}{e}' & \underset{\sim}{ab}' & \underset{\sim}{a}' & abK & eKT & aKT^2 \\
\underset{\sim}{e}' & \underset{\sim}{a}' & \underset{\sim}{b}' & bK & aKT^2 & eKT
\end{array}
$$

where $\underset{\sim}{a} = [a \; a \; a]$ and $\underset{\sim}{a}' = \begin{bmatrix} a \\ a \\ a \end{bmatrix}$. Explicitly

$$G = \begin{bmatrix}
e & e & e & e & e & e & e & e & e & e & e & e \\
e & e & e & b & b & b & ab & ab & ab & a & a & a \\
e & e & e & ab & ab & ab & a & a & a & b & b & b \\
e & ab & b & a & b & ab & b & a & e & ab & e & a \\
e & ab & b & ab & a & b & e & b & a & a & ab & e \\
e & ab & b & b & ab & a & a & e & b & e & a & ab \\
e & b & a & ab & e & a & b & e & ab & b & ab & a \\
e & b & a & a & ab & e & ab & b & e & a & b & ab \\
e & b & a & e & a & ab & e & ab & b & ab & a & b \\
e & a & ab & b & a & e & b & ab & a & b & e & ab \\
e & a & ab & e & b & a & a & b & ab & ab & b & e \\
e & a & ab & a & e & b & ab & a & b & e & ab & b
\end{bmatrix}$$

is a $GH(12, Z_2 \times Z_2)$.

4. USING GW(v,k,G) TO CONSTRUCT SBIBD

Write P for the matrix with 1 where David Glynn's $GW(13,9,S_3)$ has zeros and 0 where the GW is non-zero and $e = (1,1,1,1,1,1)$. Then, as Glynn observed,

$$DG = \begin{bmatrix}
P^T & I_{13} \times e \\
I_{13} \times e^T & \begin{array}{l} GW(13,9,S_3) \text{ with the} \\ \text{group elements replaced} \\ \text{by their permutation} \\ \text{matrix representation} \end{array}
\end{bmatrix}$$

is the incidence matrix of the Hughes plane of order 9.

In general, we can say

Lemma 3. *Suppose there exists a* $GW(p^2+p+1, p^2, G)$, $|G| = p(p-1)$. *Then forming DG similarly to the above we have the incidence matrix of a tangentially transitive projective plane of order* p^2.

Remark. If G is an "interesting group" then the related projective plane will also be "interesting".

We now give some other constructions using generalised weighing matrices.

Theorem 4. *Suppose there is a generalised balanced weighing matrix* $W = GW(v,k,Z_d)$ *with entries,* θ^i, *which are* d^{th} *roots of unity. Suppose the underlying SBIBD has parameters* (v,k,λ). *Then if* $d(v-k) = k - 1$ *there exists a BIBD*
$$\left(vd^2, vd(d+1), k(d+1), kd, k\right)$$
and an SBIBD
$$\left(vd(d+1)+1, vd+1, k\right).$$

Proof. Each entry θ^i, of the $GW(v,k,Z_d)$, is first replaced by $\theta^i GH(d,Z_d)$ where $GH(d,Z_d)$ is the generalised Hadamard matrix. Now W_B of order vd^2 with kd ones per row and column is formed by replacing each element, θ^i, by its permutation matrix representation A_i of order d. W_B has inner products $0, k, \lambda$.

W_A and W_C are formed by replacing 0 by 0_d, the $d \times d$ zero matrix, and θ^i by $e \times A_i$ and $e^T \times A_i$ respectively in W, with e the $d \times 1$ matrix of ones. Now W_A has inner products $k,0,\lambda/d$, is of size $vd^2 \times vd$, and has k ones per row and kd ones per column. $\begin{bmatrix} W_A & W_B \end{bmatrix}$ is the required BIBD.

The matrix W_D is now obtained by replacing each zero element of W by J_d the $d \times d$ matrix of ones and each non-zero element by 0_d. Then, with f the $1 \times vd$ matrix of ones,

$$\begin{bmatrix} 1 & f & 0 \\ f^T & W_D & W_C \\ 0 & W_A & W_B \end{bmatrix}$$

is the required SBIBD.

Example. Berman has shown that there is a circulant matrix $W = W\big((2^{t+1}-1)/3, 2^{t-1}\big)$, $t \geq 3$ odd, with entries the cube roots of unity $1, \omega, \omega^2$. Since 3 is prime, W is a balanced $GW\big((2^{t+1}-1)/3, 2^{t-1}, Z_3\big)$. We replace each element ω^i by

$$\omega^i \begin{bmatrix} 1 & 1 & 1 \\ 1 & \omega & \omega^2 \\ 1 & \omega^2 & \omega \end{bmatrix}$$

and 0 by 0_3. We form W_B by replacing ω^i by $\begin{bmatrix} 0 & 1 & 0 \\ 0 & 0 & 1 \\ 1 & 0 & 0 \end{bmatrix}^i$ and 0 by 0_3.

W_A and W_C are obtained by replacing 0 by 0_3 and ω^i by

$$\begin{bmatrix} 1 \\ 1 \\ 1 \end{bmatrix} \times \begin{bmatrix} 0 & 1 & 0 \\ 0 & 0 & 1 \\ 1 & 0 & 0 \end{bmatrix}^i \quad \text{and} \quad [1\ 1\ 1] \times \begin{bmatrix} 0 & 1 & 0 \\ 0 & 0 & 1 \\ 1 & 0 & 0 \end{bmatrix}^i$$

respectively.

Since W is orthogonal, the inner product of any two rows is a multiple λ of $1 + \omega + \omega^2$. Further, since replacing $1, \omega, \omega^2$ by 1 gives the incidence matrix of a $\big((2^{t+1}-1)/3, 2^{t-1}, 3.2^{t-3}\big)$ difference set, we see $\lambda = 2^{t-3}$. Now W_B is of order $3(2^{t+1}-1)$ has 3.2^{t-1} ones per row and column and has inner products of rows $0, 2^{t-1}$ or 3.2^{t-3}; W_A is of size $3(2^{t+1}-1) \times (2^{t+1}-1)$, has 2^{t-1} ones per row and $2^{t-1}, 0$ or 2^{t-3}; W_C is of size $(2^{t+1}-1) \times 3(2^{t+1}-1)$, has 3.2^{t-1} ones per row and 2^{t-1} ones per column; further, it has inner products 0 or 3.2^{t-3}.

$\begin{bmatrix} W_A & W_B \end{bmatrix}$ is a BIBD $\big(3(2^{t+1}-1),\ 2^{t+3}-4, 2^{t+1}, 3.2^{t-1}, 2^{t-1}\big)$.

We form W_D by replacing the zeros of W by J_3 and all other elements by 0_3. Since W_D is based on a $\big((2^{t+1}-3)/3, (2^{t-1}-1)/3, (2^{t-3}-1)/3\big)$ difference set, it has $2^{t-1}-1$ ones per row and column and inner products $2^{t-3}-1$ and $2^{t-1}-1$. So with e the $1 \times (2^{t+1}-1)$ matrix of ones we have

$$\begin{bmatrix} 1 & e & 0 \\ e^T & W_D & W_C \\ 0 & W_A & W_B \end{bmatrix}$$

is the incidence matrix of a $(2^{t+3}-3,2^{t+1},2^{t-1})$ SBIBD.

So we have a new proof of a case of a theorem of Radjundlia.

Corollary 5. *Let* $t > 3$ *be odd. Then there exists an SBIBD with parameters* $(2^{t+3}-3,2^{t+1},2^{t-1})$.

Example. Berman exhibits a $W(16,21)$ with entries which are cube roots of unity. Since $d = 3$, $v = 21$, $k = 16$ satisfies $3(21-16) = 16-1$, the theorem tells us there is an SBIBD $(253,64,16)$.

Corollary 6. *Suppose there is a* $GW(p+1,p,Z_{p-1})$. *Then there exists an SBIBD with parameters* $(p(p^2-1)+1,p^2,p)$. *In particular, an SBIBD* $(p(p^2-1)+1,p^2,p)$ *exists whenever* p *is a prime power.*

This family of SBIBDs has recently been found by Becker and Piper [1] and in more general form by Rajkundlia.

Theorem 7. *Suppose there is a balanced generalised weighing matrix* $GW(v,k,Z_d)$. *Suppose the underlying SBIBD has parameters* (v,k,λ). *Then if* $v - 1 = (v-k)(d-1)$ *there exists an SBIBD*
$$\bigl(dv,k+d(v-k),d(v-k)\bigr).$$

Proof. Replace each non-zero element by its $d \times d$ permutation matrix representation and each zero element by the $d \times d$ matrix of ones.

Berman found circulant $W\bigl((2^{t+1}-1)/3,2^{t-1}\bigr)$, $t \geq 3$ odd, with entries which are cube roots of unity. Since 3 is a prime, this matrix is a balanced $GW\bigl((2^{t+1}-1)/3,2^{t-1},Z_3\bigr)$. This satisfies the conditions of the theorem and so we have the family of SBIBDs $(2^{t+1}-1,2^t-1,2^{t-1}-1)$ which is, of course, well-known.

Corollary 8. *Suppose there exists a* $GW(p^2+1,p^2,Z_{p+1})$. *Then there exists an* SBIBD $\left(\dfrac{p^4-1}{p-1}, \dfrac{p^3-1}{p-1}, \dfrac{p^2-1}{p-1}\right)$.

This gives the well-known family of SBIBDs $\left(\dfrac{p^4-1}{p-1}, \dfrac{p^3-1}{p-1}, \dfrac{p^2-1}{p-1}\right)$ when p is a prime power for in this case we know the $GW(p^2+1,p^2,Z_{p+1})$ exists from Theorem 1.

5. USING GENERALISED HADAMARD MATRICES

We now give an alternate construction for the SBIBD of Corollary 6.

Theorem 9. *Suppose there exists a generalised Hadamard matrix* $GH\bigl(qp^i(p-1),C_p\bigr)$ *where* C_p *is an abelian group. Further, suppose an SBIBD* $\bigl(p(qp^i-1,qp^i,qp^i-1)\bigr)$ *exists with incidence matrix containing* $M_1 = J_{qp^i-p+1}$. *Then there exists an SBIBD* $\bigl(p(qp^{i+1}-1)+1,qp^{i+1},qp^i\bigr)$.

Proof. Let e_t be the $1 \times t$ matrix of ones and J_t the $t \times t$ matrix of ones. Let 0_a, 0_b and 0_c be zero matrices of sizes $x \times y$, $y \times x$ and $y \times y$ respectively, where $x = p(qp^i-p+1)$ and $y = p^2 - 2p + 1$. Let $A_1,\dots,A_p$ be the $p \times p$ permutation matrix representation of C_p. Write GH(A) for the (0,1) matrix obtained by replacing each element of C_p by its appropriate matrix representation. Then GH(A) is a symmetrical group divisible design with parameters

$$\left(qp^{i+1}(p-1),qp^{i+1}(p-1),qp^i(p-1),qp^i(p-1),0,qp^{i-1}(p-1),qp^i(p-1),p\right).$$

We write the incidence matrix of the SBIBD $\left(p(qp^i-1)+1,qp^i,qp^{i-1}\right)$ as

$$\begin{bmatrix} M_1 & M_2 \\ M_3 & M_4 \end{bmatrix} = \begin{bmatrix} M_1 & X \\ M_3 & \end{bmatrix} = \begin{bmatrix} M_1 & M_2 \\ Y & \end{bmatrix} \quad ,$$

where M_1 is $(qp^i-p+1) \times (qp^i-p+1)$, M_2 is $(qp^i-p+1) \times qp^i(p-1)$, M_3 is $qp^i(p-1) \times (qp^i-p+1)$ and M_4 is $qp^i(p-1) \times qp^i(p-1)$. Now form

$$M = \left[\begin{array}{cc|c} M_1 \times J_p & 0_a & \\ & & e_p \times X \\ 0_b & 0_c & \\ \hline & e_p^T \times Y & GH(A) \end{array} \right]$$

which is the incidence matrix of the required SBIBD.

We note in passing that

$$[e_p^T \times Y \quad GH(A)]$$

is a pairwise balanced design $\left(qp^{i+1}(p-1);qp^{i+1},qp^i(p-1);qp^i\right)$.

In particular, we note that if $q = 1$ and p and p - 1 are both prime powers, the $GH\left(p^i(p-1),C_p\right)$ exists for all positive i, as does the SBIBD $(p^2-p+1,p,1)$. So an SBIBD $\left(p(p^2-1)+1,p^2,p\right)$ of the right form exists by the theorem. Hence, by induction we have Rajkundlia's theorem as a corollary.

Corollary 9. *Suppose p and p - 1 are prime powers. Then there exists an SBIBD $\left(p(p^{i+1}-1)+1,p^{i+1},p^i\right)$ for all positive i.*

Example.

$$
\left[
\begin{array}{ccc|cccccc}
1 & 1 & 1 & e & e & & & & \\
1 & 1 & 1 & & & e & e & & \\
1 & 1 & 1 & & & & & e & e \\
 & & & e & & e & & & \\
 & & & e & & & & e & e \\
 & & & & e & e & & & e \\
 & & & & e & & e & e & \\
\hline
f & & f & f & I & I & I & I & I & I \\
f & & & f & f & I & I & T & T^2 & T^2 & T \\
 & f & & f & & f & I & T & I & T & T^2 & T^2 \\
 & f & & & f & & f & I & T^2 & T & I & T & T^2 \\
 & & f & f & & & f & I & T^2 & T^2 & T & I & T \\
 & & f & & f & f & I & T & T^2 & T^2 & T & I \\
\end{array}
\right]
$$

where $e = [1\ 1\ 1]$ and $f = \begin{bmatrix} 1 \\ 1 \\ 1 \end{bmatrix}$ is a $(25,9,3)$.

REFERENCES

[1]   H. Beker and F. Piper, Some designs which admit strong tactical decomposi-
      tions, *J. Combinatorial Theory*, Series A, 22 (1977), 38-42.

[2]   Gerald Berman, Weighing matrices and group divisible designs determined by
      EG(t,p^n), t > 2, *Utilitas Math.* 12 (1977), 183-192.

[3]   Gerald Berman, Families of generalised weighing matrices, CODR 77/18, Dept.
      Combinatorics and Optimisation, Univ. of Waterloo, Waterloo, Canada   1977.
      Canad. J. Math. 30 (1978), 1016-1028.

[4]   A.T. Butson, Generalised Hadamard matrices, *Proc. Amer. Math. Soc.* 13 (1962)
      894-898.

[5]   A.T. Butson, Relations among generalised Hadamard matrices, relative
      difference sets and maximal length recurring sequences, *Canad. J. Math.* 15
      (1963), 42-48.

[6]   P. Delsarte and J.M. Goethals, Tri-weight codes and generalised Hadamard
      matrices, *Information and Control*, 15 (1969), 196-206.

[7]   P. Delsarte and J.M. Goethals, On quadratic residue-like sequences in
      Abelian groups, Report R168, MBLE Research Laboratory, Brussels, 1971.

[8]   David A. Drake, Partial λ-geometries and generalised Hadamard matrices over
      groups, *Canad. J. Math.* (to appear).

[9]   Anthony V. Geramita and Joan Murphy Geramita, Complex orthogonal designs,
      J. Combinatorial Theory, Series A.

[10] Anthony V. Geramita and Jennifer Seberry, *Orthogonal designs: Quadratic
      Forms and Hadamard Matrices*, Marcel Dekker, New York, 1978.

[11]   Anthony V. Geramita, Norman J. Pullman and Jennifer Seberry Wallis,
         Families of weighing matrices, *Bull. Austral. Math. Soc.* 10 (1974), 119-122
         (= Report of Algebra Group, Queen's University, Kingston 1972/3, (1973),
         260-264).

[12]   David Glynn, *Finite projective planes and related combinatorial systems*,
         Ph.D. Thesis, University of Adelaide, Adelaide, 1978.

[13]   Richard Hain, *Circulant weighing matrices*, M.Sc. Thesis, A.N.U., Canberra
         1977.

[14]   R.C. Mullin, A note on balanced weighing matrices. *Combinatorial Mathe-
         matics III*, Proceedings of Third Australian Conference on Combinatorial
         Mathematics, Brisbane, 1974, in Lecture Notes in Mathematics, Vol. 452,
         Springer-Verlag, Berlin-Heidelberg-New York, 1975, 28-41.

[15]   R.C. Mullin and R.G. Stanton, Group matrices and balanced weighing designs,
         Utilitas Math. 8 (1975), 277-301.

[16]   R.C. Mullin and R.G. Stanton, Balanced weighing matrices and group
         divisible designs, *Utilitas Math.* 8 (1975), 303-310.

[17]   Jane W. di Paola, Jennifer Seberry Wallis and W.D. Wallis, A list of
         (v,b,r,k,λ) designs for $r \leqslant 30$. *Proc. Fourth Southeastern Conference on
         Combinatorics, Graph Theory and Computing*, Congressus Numeratium Utilitas
         Publishing Corp., Winnipeg, Canada, 1973, 249-258.

[18]   Dinesh Rajkundlia, *Some techniques for constructing new infinite families
         of balanced incomplete block designs*, Ph.D. Dissertation, Queen's Univer-
         sity, Kingston, Canada, 1978 (= Queen's Mathematical Reprint No. 1978-16).

[19]   Jennifer Seberry Wallis, Hadamard matrices. Part IV of *Combinatorics:
         Room Sequences, sum free sets and Hadamard matrices* by W.D. Wallis, Anne
         Penfold Street and Jennifer Seberry Wallis, in Lecture Notes in Mathe-
         matics, Vol. 292, Springer-Verlag, Berlin-Heidelberg-New York, 1972,
         273-489.

[20]   S.S. Shrikhande, Generalised Hadamard matrices and orthogonal arrays of
         strength two, *Canad. J. Math.* 16 (1964), 736-740.

[21]   R.G. Stanton and R.C. Mullin, On the non-existence of a class of circulant
         balanced weighing matrices, SIAM, *J. Appl. Math.* 30 (1976), 98-102.

[22]   G.A. Vanstone and R.C. Mullin, A note on existence of weighing matrices
         $W(2^{2n-j}, 2^n)$ and associated combinatorial designs, *Utilitas Math.* 8 (1975),
         371-381.

[23]   F. Yates, Complex experiments, *J. Royal Stat. Soc.* B2 (1935), 181-223.

Department of Applied Mathematics,
University of Sydney,
Sydney, Australia.

BALANCED BINARY ARRAYS I: THE SQUARE GRID

ANNE PENFOLD STREET AND SHEILA OATES MACDONALD

ABSTRACT

We consider the following problem arising in agricultural statistics. Suppose that a large number of plants are set out on a regular grid, which may be triangular, square or hexagonal, and that among these plants, half are to be given one and half the other of two possible treatments. For the sake of statistical balance, we require also that, if one plant in every k plants has i of its immediate neighbours receiving the same treatment as itself, then k is constant over all possible values of i. Such arrays are constructed for the square grid, and special properties are considered which may simplify their application.

STATEMENT OF THE PROBLEM

Plants grown in large numbers for commercial purposes are usually set out on a regular grid, triangular, square or hexagonal. In order to space them for optimal yield, it is necessary to study the way they behave when competing with each other for light.

Consider the experimental arrangements in a greenhouse. Over each plant there is a lamp; half the plants are to have their lamps on and half off. The intensity of light falling on any particular plant is determined firstly by whether its own lamp is on and secondly by the number of its immediate neighbours whose lamps are on. (We may ignore more distant lamps to within a reasonable approximation.)

For example, in the square grid, each plant has four immediate neighbours. Thus each plant can lie at the centre of any of $2^4 = 16$ possible neighbourhoods, and these 16 neighbourhoods fall into five equivalence classes depending on how many of the neighbouring plants have their lamps in the same state as the central lamp.

Suppose that 0 denotes a plant whose lamp is on, and 1 a plant whose lamp is off. Figure 1 shows the 16 possible neighbourhoods of 0, grouped into five equivalence classes. For convenience in later discussion, we show the class of γ configurations subdivided into γ_1 and γ_2, according to the way the 1s adjacent to the central 0 are placed relative to each other. Note that we always label the configurations according to the number of neighbours which match the central element; thus 1 is in the α configuration when surrounded by 1s.

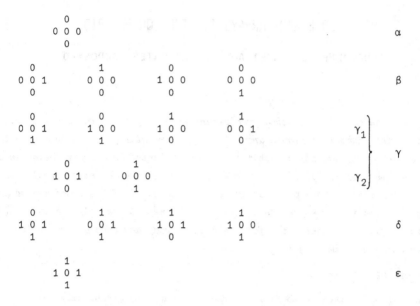

Figure 1: the neighbourhoods of 0.

For the sake of statistical balance, the array is required to consist of equal numbers of 0s and 1s and, further, to have equal numbers of each symbol in each configuration. Hence on average, of every ten elements in the array, we should have five 0s and five 1s, and of the five symbols of each type, one should be in each of the α, β, γ, δ and ε configurations.

When the rows of such an array are to be marked out consecutively in an experiment, it is desirable for the array to be *sequential*: that is, for each of its rows and columns to be occupied by the same periodic binary sequence which may always run in the same direction or may reverse. For convenience, the period of this sequence should be short. If sequentiality is not possible, then some other simplifying property is desirable.

If the plants are placed on a triangular grid, each lies at the centre of one of $2^3 = 8$ possible neighbourhoods, which fall into four equivalence classes. Similarly, there are $2^6 = 64$ neighbourhoods in the hexagonal grid and these fall into seven classes. In either case, a sequential array has the same sequence along its rows in all three directions.

We heard of this problem originally from Professor Richard Cormack (St.Andrew's University) [1] and have since discussed it and other closely related problems with Dr. David Gates (DMS, CSIRO) [2]. In this paper we consider balanced arrays for the square grid.

SEQUENTIAL ARRAYS FOR THE SQUARE GRID

Consider a balanced sequential array of minimum period. Any 0 in the sequence must occupy one of three positions: between two 0s, which we shall denote by M for middle, between a 0 and a 1 (E for end) or between two 1s (I for isolate). Table 1 shows the average contributions of each type of 0 to a sequence by each of the configurations of Figure 1.

	M	E	I
α	1	0	0
β	½	½	0
γ_1	0	1	0
γ_2	½	0	½
δ	0	½	½
ε	0	0	1

Table 1: average contribution to sequence of types of 0s.

Thus a sequence containing five 0s, one each of α, β, γ_1, δ and ε per period would have (on average) 1½ type M, 2 type E and 1½ type I. To have whole numbers we must thus have two of each of α, β, γ_1, δ and ε, giving ten 0s in all. A sequence containing one each of α, β, γ_2, δ and ε would have 2 type M, 1 type E and 2 type I, but we cannot have only one end per period, so again we must double, giving ten 0s in all. We denote such sequences respectively as type (3,4,3) and type (4,2,4) sequences in 0. (Note that a sequence containing both γ_1 and γ_2 would have to contain at least twenty 0s). Similar arguments apply to the 1s, so the minimum period of the sequence must be twenty. (But notice that the sequence could be type (3,4,3) in one symbol and type (4,2,4) in the other.) Longer sequences have period 20n and type a(3,4,3) + b(4,2,4) in either symbol, where a+b = n.

A sequence of period 20 and type (4,2,4) in 0 must contain one 6-string of 0s and four isolated 0s. A sequence of period 20 and type (3,4,3) in 0 must contain three isolated 0s, together with two strings of 0s, either a 5-string and a 2-string, or a 4-string and a 3-string. The possible configurations for these 0s are shown in Figure 2.

Even for the minimum period, we still have a tremendous number of available sequences. These sequences could be enumerated by the methods of Hutchinson [5] since 0s of types M, E and I are respectively associated with length three subsequences of the forms 000, 001 (equivalently 100) and 101. However, so few of the sequences lead to balanced arrays that this approach seems to be of little use. The most obvious kind of array to start looking for is a cyclic array; we have in fact aimed at a generalisation of this, in the following sense.

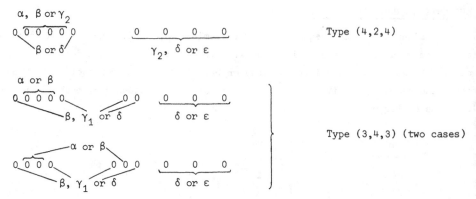

Figure 2: possible configurations for 0s in minimum period sequence.

Consider an n-string of 0s in some row of a cyclic array. As the n-string cycles in the array, its neighbours cycle with it, so that the corresponding n-strings in each row of the array appear in the same configurations. However the configurations might well remain constant without the array being cyclic.

We therefore call an array of minimum period *row-uniform* if the symbols of corresponding strings of each row are in the same configurations, and we define *column-uniform* similarly. An array might be both row- and column-uniform but the configurations of symbols in the strings of the rows need not match those of the columns. However if the configurations do match, we call the array *doubly uniform*. In a row-uniform array, each row must contain two symbols in each configuration; this restricts the possibilities shown in Figure 2 to those shown in Figure 3.

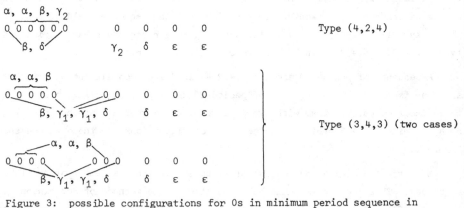

Figure 3: possible configurations for 0s in minimum period sequence in row-uniform array.

The definition of uniformity for arrays of longer period could be generalised in several ways; for the moment, we refrain from making any such definition.

Our first result shows the non-existence of certain arrays.

LEMMA 1. *No row-uniform array of period* 20 *and type* (4,2,4) *in one (or both) symbols can exist.*

Proof. There are 12 possible inequivalent configurations for the 6-string in the sequence:

(1) $\beta \alpha \alpha \beta \gamma_2 \delta$;	(7) $\beta \beta \alpha \alpha \gamma_2 \delta$;
(2) $\beta \alpha \alpha \gamma_2 \beta \delta$;	(8) $\beta \gamma_2 \alpha \alpha \beta \delta$;
(3) $\beta \alpha \beta \alpha \gamma_2 \delta$;	(9) $\beta \beta \alpha \gamma_2 \alpha \delta$;
(4) $\beta \alpha \gamma_2 \alpha \beta \delta$;	(10) $\beta \gamma_2 \alpha \beta \alpha \delta$;
(5) $\beta \alpha \beta \gamma_2 \alpha \delta$;	(11) $\beta \beta \gamma_2 \alpha \alpha \delta$;
(6) $\beta \alpha \gamma_2 \beta \alpha \delta$;	(12) $\beta \gamma_2 \beta \alpha \alpha \delta$.

Seven of these possibilities can be ruled out immediately, since they imply the existence of 2- or 3-strings of 0s, which are not possible in this sequence. These are shown in Figure 4, where the short strings are indicated by a curly bracket.

```
(4)    0 0 1 0 0 1
     1 0 0 0 0 0 0 1              (10)    0 1 0 0 0 1
       0 0 1 0 1 1                      1 0 0 0 0 0 0 1
                                          0 1 0 1 0 1
(6)    0 0 1 0 0 1
     1 0 0 0 0 0 0 1              (11)    0 0 1 0 0 1
       0 0 1 1 0 1                      1 0 0 0 0 0 0 1
                                          0 1 1 0 0 1
(7)    0 0 0 0 1 1
     1 0 0 0 0 0 0 1              (12)    0 1 0 0 0 1
       0 1 0 0 1 1                      1 0 0 0 0 0 0 1
                                          0 1 1 0 0 1
(8)    0 1 0 0 0 1
     1 0 0 0 0 0 0 1
       0 1 0 0 1 1
```

Figure 4: forbidden configurations for the 6-string in type (4,2,4) sequence of period 20.

We have five remaining configurations to consider. The proof that each of them is impossible is shown in Figure 5. The original configuration is shown in scribe (0,1) in the diagram. The sequences implied by it are shown in italics *(0,1)*; these usually involve additional strings of 0s whose neighbourhoods are determined by the uniformity condition and shown in italics also. The third stage (if needed) is shown in gothic **(0,1)**.

In (1), the column marked with an arrow contains a 7-string of 0s. Case (2) divides into two subcases: (a) leads to a 3-string of 0s in a column; (b) leads to two strings of 0s within one period of the sequence, also shown in a column. The two adjacent 1s indicated in (3) make it impossible for the 6-string below them to be in the required configuration for row-uniformity. Again in (5), the 101 sub-

(1)
```
          0 0 0 0 1 1
    1 1 1 0 0 0 0 0 0 1
  1 0 0 0 0 0 0 1 1 1
  1 1 1 0 0 0 0 0 0 1
1 0 0 0 0 0 0 1 1 1
1 1 0 0 0 0 0 0 1
    0 0 0 1 1 1
         ↑
```

(2)
```
      1   1 0 0 0
    1 0 0 0 0 0 0 1 0 1
      1 * 1 0 0 0 0 0 0 1
    1 0 0 0 0 0 0 1 1 1
      1   1 0 0 0
```

Case (a): * = 0   Case (b): * = 1

```
  1 1 1 0 0 0
1 0 0|0 0 0 0 1 0 1
  1 0|1 0 0 0 0 0 0 1
1 0 0|0 0 0 0 1 1 1
  1 1 1 0 0 0
```

```
  1 0|1 0 0 0
1 0 0|0 0 0 0 1 0 1
  1 1 1 0 0 0 0 0 0 1
1 0 0|0 0 0 0 1 1 1
  1 0|1 0 0 0
```

(3)
```
      1 1 0 1 0 0
    1 0 0 0 0 0 0 1 1
      1 1 0 0 0 0 0 0 1
1 0 0 0 0 0 0 1 0 1 1
        0 0
```

(5)
```
      1 0 1 1 0 0
    1 0 0 0 0 0 0 1 0 1
      1 0 1 0 0 0 0 0 0 1
1 0 0 0 0 0 0 1 1 0 1
```

(9)
```
      1 0 1 0 1 0
    1 0 0 0 0 0|0 1 0 1
      1 0 1 0 0|0 0 0 0 1
        0 1 0 1 0 1
```

Figure 5: configurations for the 6-string in type (4,2,4) sequence
of period 20, which are forbidden by row-uniformity.

sequence indicated makes it impossible for the 6-string below it to be in the required configuration. Finally (9) leads to a 2-string of 0s in a column.

This completes the proof of the lemma. Note that column-uniformity was not needed in the proof.

Our next result also shows non-existence of certain arrays.

LEMMA 2. *No row-uniform and column-uniform array of period 20, having a 4-string, a 3-string and three isolates of one (or both) symbols, can exist.*

Proof. There are eleven possible inequivalent configurations for the 3- and 4-strings in the sequence:

(1) $\beta\alpha\alpha\delta$ with $\gamma_1\beta\gamma_1$;   (7) $\beta\alpha\beta\gamma_1$ with $\gamma_1\alpha\delta$;

(2) $\beta\alpha\alpha\gamma_1$ " $\gamma_1\beta\delta$;   (8) $\gamma_1\alpha\beta\beta$ " $\gamma_1\alpha\delta$;

(3) $\gamma_1\alpha\alpha\delta$ " $\beta\beta\gamma_1$;   (9) $\gamma_1\alpha\beta\delta$ " $\beta\alpha\gamma_1$;

(4) $\gamma_1\alpha\alpha\gamma_1$ " $\beta\beta\delta$;   (10) $\delta\alpha\beta\gamma_1$ " $\beta\alpha\gamma_1$;

(5) $\beta\alpha\beta\delta$ " $\gamma_1\alpha\gamma_1$;   (11) $\gamma_1\alpha\beta\gamma_1$ " $\beta\alpha\delta$.

(6) $\delta\alpha\beta\beta$ " $\gamma_1\alpha\gamma_1$;

Again some cases can be ruled out immediately, though because γ_1 is not symmetrical with respect to the string, more subcases arise than in the previous lemma. The configurations are shown in Figure 6. The cases (3), (4a), (9b), (10b) and (11d) all imply the existence of 2-strings of 0s and are clearly impossible.

Cases (4b), (7a) and (10a) are rather awkward to deal with, and we shall leave them till last. The easier cases are illustrated in Figure 7 where, as in Figure 5, the original configuration is shown in scribe, the next stage in italics and the third stage (if needed) in gothic.

Case (1): If the 4-string is bordered by two 3-strings, they must both be of type (a), leading to Figure 7(i). This is impossible because the 3-strings in the columns are not uniform. If the 4-string is bordered by a 3-string (which must be of type (a)) and a 4-string, we have Figure 7(ii), which is impossible because the column 4-strings are not uniform. If the 4-string is bordered by two 4-strings, we have Figure 7(iii), which is impossible because we have 5-strings of 0s in the columns. Hence case (1) does not occur.

Case (2): Comparison of neighbourhoods shows that in neither (a) nor (b) can a 3-string border a 4-string, so we have Figure 7(iv). But this is impossible, because we have 5-strings of 0s in the marked columns. So case (2) does not occur.

Case (5): (a) If a 3-string borders a 3-string, we have Figure 7(v), which is impossible because of the 2-strings of 0s in the columns. Hence a 4-string borders the 3-string, leading (without loss of generality) to Figure 7(vi). But this also contains a 2-string of 0s in a column, ruling out case (5a).

(b) Comparison of neighbourhoods shows that the string above the 4-string must itself be a 4-string, giving Figure 7(vii). Now the column 4-string has configuration $\gamma_1\alpha\alpha\gamma_1$, so the column 3-string must have configuration $\beta\beta\delta$. But now the strings in the marked columns can be neither 3- nor 4-strings, so case (5b) and hence case (5) does not occur.

Case (6): In either (a) or (b), if the 4-string is bordered by a 4-string, we have Figure 7(viii), which is impossible because of the 2-string of 0s in the column. Hence the 4-string must be bordered by a 3-string and comparison of neighbourhoods rules out (b). Case (a) leads to Figure 7(ix), where the gothic 1s have been adjoined to prevent the 0s adjacent to them from being in the γ_2 configuration.

```
        0 0 0 1                 0 0 0               0 0 1                 1 0 1
(1)  1 0 0 0 0 1  with  (a)  1 0 0 0 1  or  (b)  1 0 0 0 1  or  (c)  1 0 0 0 1
        0 0 0 1                 1 1 1               1 1 0                 0 1 0


        0 0 0 0                 0 0 1               1 0 1
(2)  1 0 0 0 0 1  with  (a)  1 0 0 0 1  or  (b)  1 0 0 0 1
        0 0 0 1                 1 1 1               0 1 1


        0 0 0 1                 0 0 0               0 0 1
(3)  1 0 0 0 0 1  with  (a)  1 0 0 0 1  or  (b)  1 0 0 0 1
        1 0 0 1                 0 1 1               0 1 0
          ‿


              0 0 0 0                 0 0 0 1               0 0 1
(4)  (a)  1 0 0 0 0 1  or  (b)  1 0 0 0 0 1  with  1 0 0 0 1
              1 0 0 1                 1 0 0 0               0 1 1
                ‿


        0 0 0 1                 0 0 0               0 0 1
(5)  1 0 0 0 0 1  with  (a)  1 0 0 0 1  or  (b)  1 0 0 0 1
        0 0 1 1                 1 0 1               1 0 0


        1 0 0 0                 0 0 0               0 0 1
(6)  1 0 0 0 0 1  with  (a)  1 0 0 0 1  or  (b)  1 0 0 0 1
        1 0 1 0                 1 0 1               1 0 0


              0 0 0 0                 0 0 0 1               0 0 1
(7)  (a)  1 0 0 0 0 1  or  (b)  1 0 0 0 0 1  with  1 0 0 0 1
              0 0 1 1                 0 0 1 0               1 0 1


              0 0 0 0                 1 0 0 0               0 0 1
(8)  (a)  1 0 0 0 0 1  or  (b)  1 0 0 0 0 1  with  1 0 0 0 1
              1 0 1 0                 0 0 1 0               1 0 1


              0 0 0 1                 1 0 0 1               0 0 0
(9)  (a)  1 0 0 0 0 1  or  (b)  1 0 0 0 0 1  with  1 0 0 0 1
              1 0 1 1                 0 0 1 1               0 0 1


              1 0 0 0                 1 0 0 1               0 0 0
(10) (a)  1 0 0 0 0 1  or  (b)  1 0 0 0 0 1  with  1 0 0 0 1
              1 0 1 1                 1 0 1 0               0 0 1


              0 0 0 0                 0 0 0 1                 1 0 0 0
(11) (a)  1 0 0 0 0 1  or  (b)  1 0 0 0 0 1  or (c)  1 0 0 0 0 1
              1 0 1 1                 1 0 1 0                 0 0 1 1


                1 0 0 1                 0 0 1
    or (d)  1 0 0 0 0 1  with  1 0 0 0 1
                0 0 1 0                 0 0 1
```

Figure 6: configurations for the 4- and 3-strings in type (3,4,3) sequence of
                period 20.

```
 1 1 1                 1 1 1                 1 0 0 0               1 0 0 0
1 0 0 0 1             1 0 0 0 1             1 0 0 0 0 1           1 0 0 0 0 1
1 0 0 0 0 1           1 0 0 0 0 1           1 0 0 0 0 1           1 0 0 0 0 1
1 0 0 0 1             1 0 0 0 0 1           1 0 0 0 0 1           1 0 0 0 0 1
 1 1 1     (i)        1 0 0 0 1             1 0 0 0    (iii)     1 0 0 0 0 1
                       1 1 1     (ii)        ↑ ↑ ↑               0 0 0 1
                                                                  ↑ ↑ ↑     (iv)


 1 0 1                 0 0 1 1                 1                   0 1 0 1
1 0│0 0│1             1 0 0 0│0 1             1 1 0 0             1 0 0│0 0 1
1 0│0 0│1             1 0 0 0│1              1 0 0 0 0 1          1 0 0 0│0 1
 1 0 1     (v)         1 0 1     (vi)        1 0 0 0 0 1          1 0 1 0     (viii)
                                              0 0 1 1
                                               ↑ 1 ↑     (vii)


    1                     1                 1 0 1 0                   1
 1 0 1                 0 1 0 0             1 0│0 0│0 1             0 1 0 1
1 0 0 0 1             1 0 0 0 0 1          1 0│0 0│0 1            1 0 0 0 0 1
1 0 0 0 0 1           1 0 0 0 0 1           1 0 1 0    (xi)       1 0 0 0 0 1
 1 0 1 0               0 0 1 0                                     1 0 1 0
    1      (ix)         ↑ 1 ↑     (x)                              ↑ 1 ↑    (xii)


 0 1 0 0               0 0 1                 1 0 0                 1 1 0 1
1 0 0│0 0 1          1 0 0 0│1             1 0│0 0 1            1 0 0│0 0│1
1 0 0 0│0 1          1 0 0 0│0 1           1 0│0 0 0 1          1 0│0 0│0 1
 0 0 1 0   (xiii)     1 0 1 1   (xiv)       1 0 1 1   (xv)       1 0 1 1   (xvi)


 1 0 1 1               1 1 0 1               0 1 0 1               1 1 0 0
1 0│0 0│0│1          1 0│0 0 0│1          1 0 0│0 0│1           1 0 0│0 0 1
1 0│0 0│0│1          1 0│0 0 0│1          1 0│0 0 0│0 1         1 0 0 0│0 1
 1 0 1 1   (xvii)     1 0 1 1   (xviii)     1 0 1 0    (xix)     0 0 1 1     (xx)


 0 0 1 1
1 0 0 0 0│1
 1 0 0 0│0 1
  1 0 1 0   (xxi)
```

Figure 7: some forbidden configurations for strings in type (3,4,3) sequence
        of period 20.

But now the column 3-strings are not uniform, so case (6) does not occur.

Case (7b): Comparison of neighbourhoods shows that neither a 3-string nor a type
(a) 4-string can lie above a 4-string, leading to Figure 7(x), where the gothic 1s
are adjoined because the column 4-string is the longest string of 0s possible.
Hence the column 4-strings have configuration $\gamma_1 \alpha\alpha\gamma_1$, so the column 3-strings must
have configuration $\beta\beta\delta$. But now the strings in the marked columns can be neither
3- nor 4-strings, so case (7b) does not occur.

Case (8): (a) Since a 4-string, necessarily of type (a), borders the 4-string, we must have either Figure 7(xi), which is impossible since it has 2-strings of 0s in the columns, or Figure 7(xii), where the gothic 1s have been adjoined to prevent the 0s adjacent to them from being in the γ_2 configuration. So the column 3-strings have configuration $\beta\alpha\delta$, and hence the column 4-strings have configuration $\gamma_1\alpha\beta\gamma_1$. But now the strings in the marked columns can be neither 3- nor 4-strings, so case (8a) does not occur.

(b) Comparing neighbourhoods shows that the string above the 4-string must be a 4-string of type (b), leading to Figure 7(xiii), which is impossible because it has a 2-string of 0s in the column. Hence case (8) does not occur.

Case (9a): If the string above the 4-string is a 3-string, we have either Figure 7(xiv) or Figure 7(xv), both of which are impossible, since they contain 2-strings of 0s in the columns. Hence the string above the 4-string must also be a 4-string, leading to Figure 7(xvi), which is again impossible because of the 2-strings of 0s in the columns. So case (9) does not occur.

Case (11): (a) Since the 4-string is bordered by a 4-string of type (a), we have either Figure 7(xvii) or Figure 7(xviii); neither of these is possible, because they have 2-strings of 0s in the columns.

(b) and (c): In both cases, the 4-strings must be bordered by 4-strings, which could be of either type, yielding Figure 7(xix), (xx) or (xxi), each of which contains a 2-string of 0s. Hence case (11) does not occur.

We have now ruled out all but the three awkward cases, that is (4b), (7a) and (10a).

Case (4b): Comparison of neighbourhoods shows that the strings bordering the 4-string must themselves be 4-strings, and similarly that the string above the 3-string must itself be a 3-string. This leads to Figure 8, from which we see that the 4-strings must form an infinite diagonal stripe and that the 3-strings must form connected blocks containing eight 0s each. The gothic 1s in Figure 8(ii) are adjoined to prevent the 0s adjacent to them from being in the γ_2 configuration.

```
      ... 1                              1
       0 0 0 1                        1 1 0 1
    1 0 0 0 0 1                    1 0 0 0 1
       1 0 0 0 0 1                    1 0 0 0 1
         1 0 0 0 0 1                    1 0 1 1
           1 0 0 0 0 1                  1
              ...
            (i)                            (ii)
```

Figure 8: configurations of 3- and 4-strings in case (4b).

Suppose that there exists a minimal balanced row-uniform array based on this configuration. It must have period 20; it must also be doubly-uniform, as is obvious from Figure 8. The block of 3-strings can have two essentially different positions relative to the stripe of 4-strings, as shown in Figure 9.

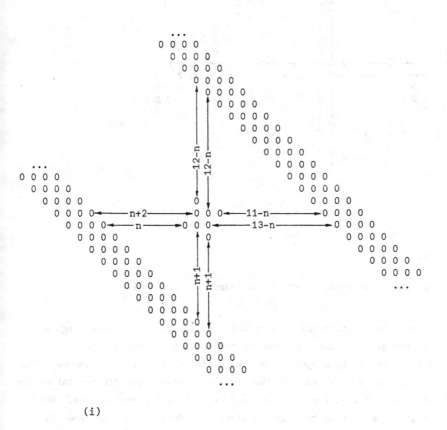

(i)

Figure 9: spacing of an array based on configuration of case (4b).

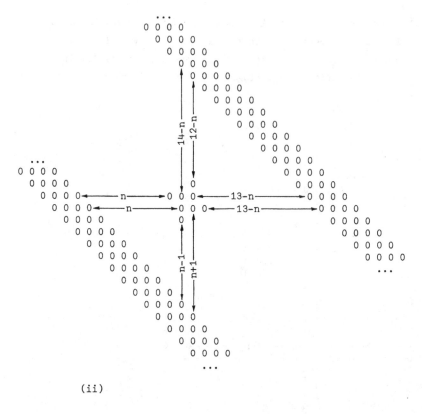

(ii)

Figure 9: spacing of an array based on configuration of case (4b).

The sequence must contain thirteen symbols besides the 3- and 4-strings of 0s;
hence, if the strings of 0s are separated by n symbols as shown in Figure 9(i),
they must be separated by 13-n symbols in the remaining part of the sequence. The
position shown in Figure 9(i) leads to the distances between the strings taking the
values n, n+2, 11-n and 13-n for the rows, and n+1, 12-n, for the columns. But
among these values, more than two must be distinct modulo 20, so the array is not
sequential. Similarly, the distances between the strings in Figure 9(ii) are n and
13-n for the rows, and n-1, n+1, 12-n and 14-n for the columns, and again more than
two of these are distinct modulo 20. This rules out case (4b).

Case (7a): For uniformity of the 4-strings, we must have either Figure 10(i),
which is impossible because of the 2-strings of 0s in the columns, or Figure 10(ii).
If the bottom string of 0s in Figure 10(ii) is a 3-string, we adjoin further symbols
in gothic to preserve uniformity giving Figure 10(iii). Now the column 3-string has
configuration $\beta\alpha\gamma_1$, and the column 4-string has configuration $\gamma_1\beta\alpha\delta$, that is, the
column configuration is that of case (10a). However if the bottom string of 0s in

Figure 10(ii) is a 4-string, we adjoin symbols in gothic to preserve uniformity, giving Figure 10(iv). Now the column 4-strings are in configuration $\beta\alpha\beta\gamma_1$, and thus the column 3-strings have configuration $\gamma_1\alpha\delta$, that is the column configuration is that of case (7a). In the case of Figure 10(iii), the 3- and 4-strings of 0s together form connected blocks, containing sixteen 0s each, whereas in the case of Figure 10(iv), the 4-strings form an infinite diagonal stripe and the 3-strings form connected blocks, containing eight 0s each.

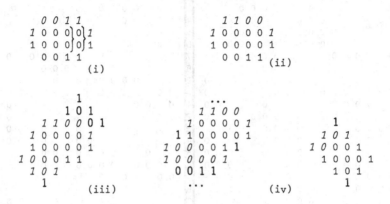

Figure 10: configurations of 3- and 4-strings in case (7a).

A balanced row- and column-uniform array based on the configuration of Figure 10(iii) is therefore not doubly uniform. It must have period 20 and its blocks of 0s may occur in the two orientations shown in Figure 11. Note that row uniformity and minimum period prevent us from rotating the block into its other possible orientations. Also, to prevent the occurrence of two 3-strings or two 4-strings in one period of the sequence, the blocks must always be stepped an even number of rows and columns from each other. We have two cases to consider: either

```
      0                   0
    0 0 0               0 0 0
  0 0 0 0             0 0 0 0
  0 0 0 0             0 0 0 0
0 0 0                     0 0 0
0                           0
      (i)        (ii)
```

Figure 11: orientations of blocks of
0s from Figure 10(iii).

all the blocks of 0s are in the same orientation, say that of Figure 11(i), or half are in each orientation.

First, suppose that all blocks are in the same orientation and that the first block lies in the first six rows and the first six columns. Then (to make sure that a 3-string occurs in each row) there must be a block occupying rows 3 to 8. In which columns can it lie? It must neither overlap nor adjoin the previous block, and hence must lie within columns 5 through 20. Now the sequentiality of the array forces the third block to be placed in the same position relative to the second as the second relative to the first. Checking modulo 20, we find that, in order to avoid two 4-strings of 0s in one period, the second block must be moved either six columns to the right, so that it occupies columns 7 to 12, or fourteen columns to the right, so that it occupies columns 15 to 20. These two possibilities are shown in Figure 12.

```
1 2 3 4 5 6 7 8 9 0 1 2 3 4 5 6 7 8 9 0        1 2 3 4 5 6 7 8 9 0 1 2 3 4 5 6 7 8 9 0
        0         0 0 0           0 0 0 0 |1|           0       0 0 0 0     0 0 0
          0 0 0         0         0 0 0 0 |2|         0 0 0     0 0 0 0       0
    0 0 0 0           0         0 0 0     |3|     0 0 0 0     0 0 0                     0
    0 0 0 0         0 0 0         0       |4|     0 0 0 0     0                       0 0 0
0 0 0         0 0 0 0             0       |5| 0 0 0                 0       0 0 0 0
0             0 0 0 0         0 0 0       |6| 0                   0 0 0     0 0 0 0
    0         0 0 0         0 0 0 0       |7|         0       0 0 0 0     0 0 0
  0 0 0         0           0 0 0 0       |8|         0 0 0     0 0 0 0       0
0 0 0           0         0 0 0         0 |9| 0       0 0 0 0     0 0 0
0 0 0         0 0 0         0           0 |0| 0 0     0 0 0 0     0                       0
0         0 0 0 0           0         0 0 |1| 0     0 0 0               0       0 0 0
0         0 0 0 0         0 0 0         0 |2| 0     0             0 0 0     0 0 0
  0 0 0           0 0 0 0             0   |3|       0       0 0 0 0     0 0 0
0 0           0           0 0 0 0         |4|       0 0 0     0 0 0 0       0
  0           0         0 0 0         0 0 0 |5|     0       0 0 0 0     0 0 0
    0 0 0         0             0 0 0       |6| 0 0 0     0 0 0 0       0
  0 0 0 0             0       0 0 0         |7| 0 0 0     0 0 0                     0       0
  0 0 0 0         0 0 0         0         |8| 0 0 0     0                       0 0 0     0
  0 0 0       0 0 0 0             0       |9| 0                 0       0 0 0 0     0 0
    0         0 0 0 0         0 0 0       |0|             0 0 0     0 0 0 0             0
```

(i) Block shifted by two rows down   (ii) Block shifted by two rows down
     and six columns right.   and fourteen columns right.

Figure 12: configurations with all blocks in orientation of Figure 11(i).

We can now read off the possible sequences from Figure 12. Neither of the arrays can be sequential. In (i), the rows require a sequence 0 0 0 0 - - - - 0 0 0 - - - 0 - - - - - and the columns a sequence 0 0 0 0 - 0 0 0 - - - 0 - - - - - - - - . Obviously these cannot agree. Similarly in (ii), the rows require a sequence 0 0 0 0 - 0 0 0 - - - - - - - - - 0 - - and the columns a sequence 0 0 0 0 - - - - 0 0 0 - - - - - 0 - - - .

We have still to consider the case where half the blocks are oriented as in Figure 11(i), and the other half as in Figure 11(ii). An argument similar to that which led to Figure 12 leads us to Figure 13, where if the first block lies in rows 1 through 6 and columns 1 through 6, the second block lies in rows 3 through 8 and either columns 7 through 12, or columns 15 through 20.

The arrays of Figure 13 are even further from sequentiality than those of Figure 12. In Figure 13(i), the row sequences include 0 0 0 0 - - - - 0 0 0 - - - 0 - - - - - , 0 0 0 0 - 0 0 0 - - - - - - - - - 0 - - , and the column sequences include 0 0 0 0 - - - 0 0 0 - - - - - - 0 - - - , and 0 0 0 0 - - 0 0 0 - - 0 - - - - - - - - . None of these sequences can agree with

each other. In Figure 13(ii), the sequences are exactly the same, although the
arrays are different.

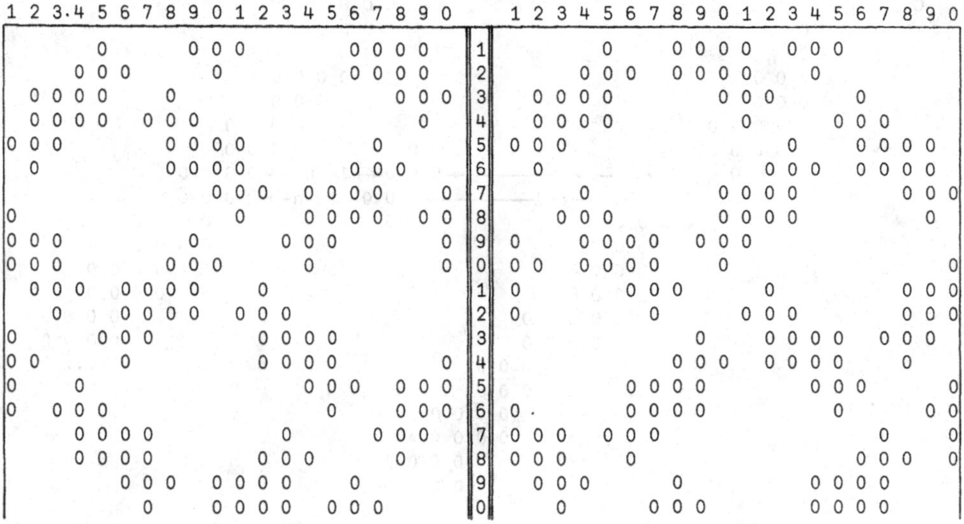

```
1 2 3.4 5 6 7 8 9 0 1 2 3 4 5 6 7 8 9 0      1 2 3 4 5 6 7 8 9 0 1 2 3 4 5 6 7 8 9 0
        0       0 0 0         0 0 0 0 |1        0       0 0 0 0     0 0 0
      0 0 0       0           0 0 0 0 |2      0 0 0     0 0 0 0       0
  0 0 0 0     0               0 0 0   |3  0 0 0 0         0 0 0         0
  0 0 0 0   0 0 0               0     |4  0 0 0 0         0         0 0 0
0 0 0       0 0 0 0           0       |5  0 0 0                 0     0 0 0 0
  0         0 0 0 0       0 0 0       |6  0                 0 0 0   0 0 0 0
          0 0 0   0 0 0 0       0     |7      0           0 0 0 0         0 0 0
0               0 0 0 0       0 0     |8    0 0 0         0 0 0 0           0
0 0 0       0         0 0 0         0 |9  0   0 0 0 0     0 0 0               0
0 0 0     0 0 0     0               0 |0  0 0 0 0 0 0       0                 0
  0 0 0   0 0 0 0     0               |1  0     0 0 0         0           0 0 0
    0   0 0 0 0     0 0 0             |2  0     0         0 0 0           0 0 0
0     0 0 0         0 0 0 0           |3          0           0 0 0 0   0 0 0
0 0     0           0 0 0 0         0 |4        0 0 0     0 0 0 0         0
0   0             0 0 0     0 0 0     |5      0 0 0 0             0 0 0         0
0   0 0 0           0       0 0 0     |6  0 .   0 0 0 0         0           0 0
  0 0 0 0     0           0 0 0       |7  0 0 0   0 0 0               0         0
  0 0 0 0       0 0 0       0         |8  0 0 0   0           0 0 0         0
    0 0 0   0 0 0 0     0             |9  0 0 0       0           0 0 0 0
    0       0 0 0 0   0 0 0           |0    0           0 0 0     0 0 0 0
```

(i) Block shifted two rows down   (ii) Block shifted two rows down
     and six columns right.   and fourteen columns right.

Figure 13: configurations with half the blocks in orientation of Figure 11(i),
            half in orientation of Figure 11(ii).

We have now shown that the configuration of Figure 10(iii) cannot lead to a
sequential array; we have still to consider that of Figure 10(iv). Here the 4-
strings of 0s form an infinite diagonal stripe and the 3-strings form connected
blocks of eight 0s each, which may occur in two possible orientations. This leads
to four possibilities, as shown in Figure 14.

The arrays of Figure 14(i), (ii) and (iii) can all be ruled out by arguments
similar to those applied to Figure 9. In (i) for instance, if the block of 3-
strings has its rows aligned with two 4-strings that fall into the same columns,
then the spaces between the 3- and 4-strings must be of lengths n, 13-n, n+1 and
12-n as shown. For the array to be sequential, these numbers can take on at most
two distinct values, forcing n = 6. But now however we place the sequence, the
blocks consisting of 3-strings of 0s overlap or adjoin, leaving no 0s in the δ
configuration. In (ii), where the block of 3-strings has its rows aligned with
two 4-strings which are not level with each other, the numbers n, 13-n, n-1 and
14-n must take at most two different values, forcing n = 7, with the same conseq-
uences as before. In (iii), the block of 3-strings is reflected to the opposite
orientation, and has its rows aligned with two 4-strings which are not level with
each other; since now n, 13-n, n-3 and 16-n are allowed only two distinct values,

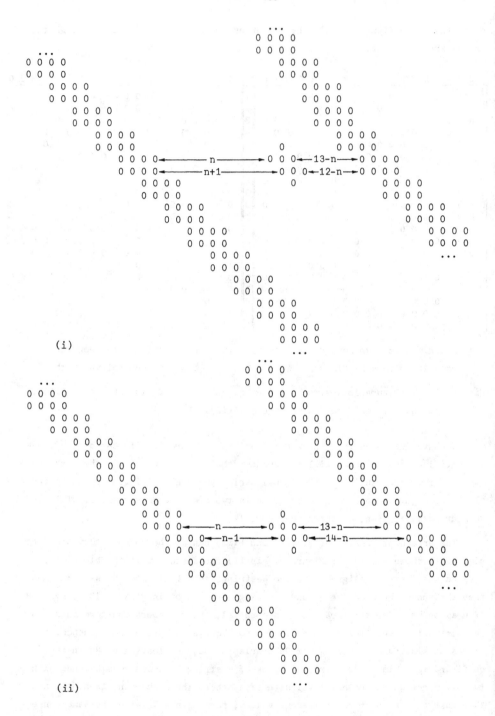

Figure 14: configurations derived from those of Figure 10(iv).

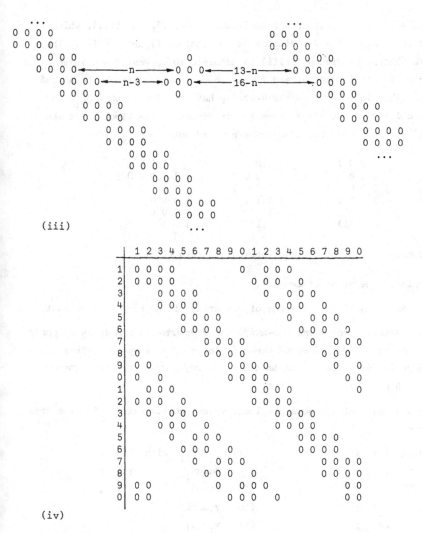

```
    ...                              ...
0 0 0 0                      0 0 0 0
0 0 0 0                      0 0 0 0
  0 0 0 0           0           0 0 0 0
  0 0 0 0◄──────n──────►0 0 0◄──13-n──►0 0 0
    0 0 0 0◄──n-3──►0 0 0◄─────16-n─────►0 0 0
    0 0 0 0           0               0 0 0
      0 0 0 0                           0 0 0 0
      0 0 0 0                           0 0 0 0
        0 0 0 0                           0 0 0 0
        0 0 0 0                           0 0 0 0
          0 0 0 0                           ...
          0 0 0 0
            0 0 0 0
            0 0 0 0
              0 0 0 0
              0 0 0 0
                ...
```

(iii)

```
   |1 2 3 4 5 6 7 8 9 0 1 2 3 4 5 6 7 8 9 0
 1 |0 0 0 0           0   0 0 0
 2 |0 0 0 0               0 0 0   0
 3 |    0 0 0 0             0   0 0 0
 4 |    0 0 0 0             0 0 0   0
 5 |      0 0 0 0             0   0 0 0
 6 |      0 0 0 0               0 0 0   0
 7 |        0 0 0 0               0   0 0 0
 8 |0       0 0 0 0               0 0 0
 9 |0 0         0 0 0 0               0   0
 0 |0   0       0 0 0 0                 0 0
 1 |  0 0 0         0 0 0 0               0
 2 |0 0 0   0       0 0 0 0
 3 |  0   0 0 0       0 0 0 0
 4 |    0 0 0   0     0 0 0 0
 5 |    0   0 0 0       0 0 0 0
 6 |      0 0 0   0     0 0 0 0
 7 |        0   0 0 0     0 0 0 0
 8 |        0 0 0   0     0 0 0 0
 9 |0 0             0   0 0 0       0 0
 0 |0 0             0 0 0   0       0 0
```

(iv)

Figure 14: configurations derived from those of Figure 10(iv).

we must have n = 8, and again the blocks of 3-strings will not fit into the array.

The case shown in Figure 14(iv) has to be considered more carefully. Here the rows of the block of 3-strings are aligned as in (i) but the block is oriented as in (iii). But the array has sequence 0 0 0 0 - - - - - 0 - 0 0 0 - - - - - - in every row and sequence 0 0 0 0 - - - - - - 0 - 0 0 0 - - - - - in every column. Hence it is not sequential and we have ruled out case (7a).

Case (10a): If the string above the 4-string (see Figure 6) is itself a 4-string, then row-uniformity forces the italic symbols in Figure 15(i), which is impossible since it has a 2-string of 0s in a column. Hence the string above the 4-string is a 3-string and we must have either the configuration of Figure 15(ii), which is again impossible because of a 2-string of 0s, or that of Figure 15(iii). The gothic 1 at the bottom of Figure 15(iii) is adjoined to prevent the 0 adjacent to it from being in γ_2 configuration, and forces the column 3-string to be a $\gamma_1\alpha\delta$ configuration. This in turn makes the 4-string have configuration $\beta\alpha\beta\gamma_1$, and leads to Figure 15(iv). But now we have the transpose of the block of 0s shown in Figure 10(iii) and this case has already been ruled out.

```
                                                     1
   1 1 0 1          0 0 1          1 0 0        1 1 0 1
  1 0 0 0 0 1      1 0 0 0 1      1 0 0 0 1    1 0 0 0 0 1
1 0 0 0 0 1      1 0 0 0 0 1    1 0 0 0 0 1      1 0 0 0 1
  1 0 1 1          1 0 1 1        1 0 1 1      1 0 0 0 1
                                      1      1 0 0 0 0 1
    (i)              (ii)          (iii)      1 0 1 1
                                               1   (iv)
```

Figure 15: configurations of 3- and 4-strings in case (10a).

This completes the proof of Lemma 2.

Our last lemma shows the existence of one array of the kind that we want.

LEMMA 3. *There exists exactly one row-uniform and column-uniform array of period 20, having a 5-string, a 2-string and three isolates of each symbol. This array is the one shown in Figure 28; it is unique up to reflection and has sequence*
0 0 0 0 0 1 0 1 0 0 1 1 1 1 1 0 1 0 1 1.

Proof. There are again eleven possible inequivalent configurations for the strings of the sequence:

(1) $\beta\alpha\alpha\beta\delta$ with $\gamma_1\gamma_1$;   (7) $\gamma_1\alpha\alpha\beta\gamma_1$ with $\beta\delta$;

(2) $\delta\alpha\alpha\beta\beta$ " $\gamma_1\gamma_1$;   (8) $\beta\alpha\beta\alpha\gamma_1$ " $\gamma_1\delta$;

(3) $\beta\alpha\alpha\beta\gamma_1$ " $\gamma_1\delta$;   (9) $\beta\alpha\beta\alpha\delta$ " $\gamma_1\gamma_1$;

(4) $\gamma_1\alpha\alpha\beta\beta$ " $\gamma_1\delta$;   (10) $\gamma_1\alpha\beta\alpha\delta$ " $\beta\gamma_1$;

(5) $\gamma_1\alpha\alpha\beta\delta$ " $\beta\gamma_1$;   (11) $\gamma_1\alpha\beta\alpha\gamma$ " $\beta\delta$.

(6) $\delta\alpha\alpha\beta\gamma_1$ " $\beta\gamma_1$;

The configurations are shown in Figure 16. The cases (5b), (6b), (7d), (10b) and (11c) imply the existence of 3-strings of 0s and can be ruled out immediately.

Several additional cases can be ruled out by comparing the neighbourhood of the 2-string with that of a 2-string bordering the 5-string, namely (2b), (4a), (7a), (7b) and (8b), the last because one of the two strings bordering the 5-string must be a 2-string. Of the remaining cases, only (5a), (9a) and (9b) are awkward to deal with and again we leave them till last.

Case (1): Both (a) and (b) are impossible, because of the configuration of the 5-string. As shown in Figure 17(i), it implies 7-strings of 0s in the columns.

```
         0 0 0 0 1                      1 1                    1 0
(1)  1 0 0 0 0 0 1  with  (a)  1 0 0 1  or  (b)  1 0 0 1
         0 0 0 1 1                      0 0                    0 1


         1 0 0 0 0                      1 1                    1 0
(2)  1 0 0 0 0 0 1  with  (a)  1 0 0 1  or  (b)  1 0 0 1
         1 0 0 1 0                      0 0                    0 1


            0 0 0 0 0                   0 0 0 0 1                0 1
(3)  (a)  1 0 0 0 0 0 1  or  (b)  1 0 0 0 0 0 1  with  1 0 0 1
            0 0 0 1 1                   0 0 0 1 0                1 1


            0 0 0 0 0                   1 0 0 0 0                0 1
(4)  (a)  1 0 0 0 0 0 1  or  (b)  1 0 0 0 0 0 1  with  1 0 0 1
            1 0 0 1 0                   0 0 0 1 0                1 1


            0 0 0 0 1                   0 0 0 1 1                0 0
(5)  (a)  1 0 0 0 0 0 1  or  (b)  1 0 0 0 0 0 1  with  1 0 0 1
            1 0 0 1 1                   1 0 0 1                  0 1
                                          ‿

            1 0 0 0 0                   1 0 0 0 1                0 0
(6)  (a)  1 0 0 0 0 0 1  or  (b)  1 0 0 0 0 0 1  with  1 0 0 1
            1 0 0 1 1                   1 0 0 1 0                0 1


            0 0 0 0 0                   0 0 0 0 1                   1 0 0 0 0
(7)  (a)  1 0 0 0 0 0 1  or  (b)  1 0 0 0 0 0 1  or  (c)  1 0 0 0 0 0 1
            1 0 0 1 1                   1 0 0 1 0                   0 0 0 1 1

                      ⌢
            1 0 0 0 1                0 1
or  (d)  1 0 0 0 0 0 1  with  1 0 0 1
            0 0 0 1 0                0 1


            0 0 0 0 0                   0 0 0 0 1                0 1
(8)  (a)  1 0 0 0 0 0 1  or  (b)  1 0 0 0 0 0 1  with  1 0 0 1
            0 0 1 0 1                   0 0 1 0 0                1 1


         0 0 0 0 1               0 0                    0 1
(9)  1 0 0 0 0 0 1  with  (a)  1 0 0 1 or  (b)  1 0 0 1
         0 0 1 0 1               1 1                    1 0


            0 0 0 0 1                1 0 0 0 1                0 0
(10) (a)  1 0 0 0 0 0 1  or  (b)  1 0 0 0 0 0 1 with  1 0 0 1
            1 0 1 0 1                0 0 1 0 1                0 1


            0 0 0 0 0                   1 0 0 0 0                   1 0 0 0 1
(11) (a)  1 0 0 0 0 0 1  or  (b)  1 0 0 0 0 0 1  or  (c)  1 0 0 0 0 0 1
            1 0 1 0 1                   0 0 1 0 1                   0 0 1 0 0

              0 1
with  1 0 0 1
              0 1
```

Figure 16: configurations of the 5- and 2-strings in type (3,4,3) sequence
of period 20.

Case (2a): Uniformity implies the existence of 3-strings of 0s in the columns; see Figure 17(ii).

Case (3): (a) Since only 5-strings and 2-strings of 0s occur, we have Figure 17(iii); the top 5-string is necessarily of type (a), yielding Figure 17(iv) or the scribe and italic symbols of Figure 17(v). Now (iv) can be ruled out immediately as the column 2-strings are non-uniform. Inspection of the strings of 0s in the columns of (v) shows that the column 5-string must be $\beta\alpha\alpha\beta\gamma_1$. Extending the two right-hand columns in accordance with this pattern yields the gothic symbols of (v). But now the marked column involves the configuration $\alpha\alpha\gamma_1$ and so column uniformity is lost.

(b) Uniformity (and no 5-strings of type (a)) implies the scribe and italic symbols of Figure 17(vi); avoiding 6-strings of 0s implies the gothic symbols which in turn give non-uniform row 5-strings.

Case (4b): Both strings bordering the 5-string must be 5-strings. Row uniformity implies Figure 17(vii), which is impossible because it has 3-strings of 0s in the columns.

Case (6a): Completing the 5-string which borders the 5-string and making it row-uniform gives the scribe and italic symbols of Figure 17(viii). The 3-strings of 0s in the columns must now be extended to 5-strings, which puts the row 2-strings into the wrong configurations.

Case (7c): Completing the 5-strings gives the scribe symbols in Figure 17(ix). Row-uniformity leads to the italic symbols and hence to a contradiction, because it implies 3-strings of 0s in the columns.

Case (8a): Comparison of neighbourhoods shows that the string below the 5-string must also be a 5-string. Completing the 5-strings and their neighbourhoods gives us the scribe and italic symbols in Figure 17(x). The gothic symbols are now implied by uniformity, to avoid γ_2 configurations and to prevent strings of 0s longer than 5-strings in the columns. The column 5-strings are now in case (6a), which is already ruled out.

Case (10a): Completing the 5-string which borders the 5-string and preventing the γ_2 configuration gives us the scribe and italic symbols of Figure 17(xi). Uniformity now implies the gothic symbols which lead to column 3-strings of 0s.

Case (11): (a) Uniformity of the upper 5-string which must be of type (a) and prevention of the γ_2 configuration for the 0s below the 5-string gives the scribe and italic symbols of Figure 17(xii). To prevent the 0s above the 5-strings from being in the γ_2 configuration we must adjoin the gothic 1s, giving column 4-strings and hence a contradiction.

```
  1 0 0 0 0 0 1
   1 1 0 0 0 0 0 1                    0 1 0 0 1
  1 0 0 0 0 0 1 1                 1 0 0│0 0 0 1          1 0 0 0 0 0 1
   1 1 0 0 0 0 0 1                1 0 0 0}0 0 1           1 1 0 0 0 0 0 1
  1 0 0 0 0 0 1 1                 1 0 0│1 0               1 0 0 0 0 0 1 1
   1 0 0 0 0 0 1                  1 1
1 0 0 0 0 0 1 1                         (ii)                      (iii)
        ↑
              (i)


    0 0 0 1 1              1 0 0                1 1
   1 0 0 0 0 0 1          1 1 0 0 0 0       0 1 0 0 0 0 0 1
   1 0 0 0 0 0 1          1 0 0 0 0 0 1     1 0 0 0 0 0 1
1 0 0 0 0 0 1 1           1 0 0 0 0 0 1     0 1 0 0 0 0 0 1
         (iv)         1 0 0 0 0 0 1 1       1 0 0 0 0 0 1 0
                               (v)          1 0 0 0 0 1
                                                1 1   (vi)


                             1 1
   0 1 0 0 0                  0 0                     1 1 0 0 0 0 0 1
  1 0 0│0 0 0 1               0 0                    1 0 0│0 0 0 1
 0 1 0 0 0}0 0 1          1 1 0 0 1                  1 1 0 0 0}0 0 1
1 0 0 0 0 0│1 0         1 0 0 0 0 0 1               1 0 0 0 0 0│1 1
 0 0 0 0 1             1 0 0 0 0 0 1               1 0 0 0 0 0 1
         (vii)         1 0 0 1 1                            (ix)
                             0 0
                             0 0
                             1 1   (viii)


              1
            1 0
           1 0 1 0 0                          1   1
    1 1 0 0 0 0 0 1       1 0 1 0 1         1 0│1 0│1
   1 0 1 0 0 0 0 0 1    1 0 0 0│0 0│1      1 0 0│0 0│0 1
  1 0 0 0 0 0 1 0 1     1 0 0}0 0}0 1      1 0 0│0 0│0 1
  1 0 0 0 0 0 1         1 0│1 0│1          1 0│1 0│1
  0 0 1 0 1              1    1              1   1
   0   1     (x)             (xi)                  (xii)

                                           1 0 1 0 0
                                        1 0 0 0│0 0 1
                                       1 0 0 0 0}0 1
                                        0 0 1 0│1
                                             1     (xiii)
```

Figure 17: some forbidden configurations for
strings in type (3,4,3) sequence
of period 20.

Case (11): (b) The italic symbols of Figure 17(xiii) are needed to complete the
upper 5-string, which must be of type (b), to make it uniform and to prevent the
γ_2 configuration for 0. This implies a column 3-string of 0s and hence a contradiction.

We now have only cases (5a), (9a) and (9b) still to deal with.

Case (9): The string above the 5-string must itself be a 5-string. Completing it
and its neighbourhood and preventing γ_2 configurations of 0s gives us the scribe
and italic symbols of Figure 18(i). Continuing in this way leads to the diagonal
stripe shown in Figure 18(ii). This applies to both cases (a) and (b). However

in case (a) we must have the 2-strings in blocks as shown in Figure 18(iii),
whereas in case (b), the 2-strings must also form a diagonal stripe as in Figure
18(iv).

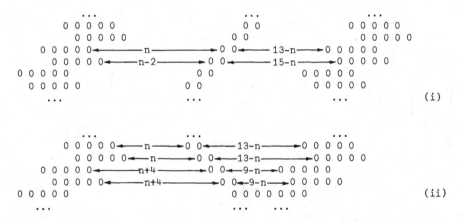

Figure 18: configurations of 5- and 2-strings in case (9).

Case (9b) can now be ruled out fairly quickly. Two possibilities have to be
considered, as shown in Figure 19.

Figure 19: configurations arising in case (9b).

For the array of Figure 19(i) to be sequential, we must have only two distinct
values among n, n-2, 13-n and 15-n, where n is the number of spaces in the sequence
between the 5-string and the 2-string. But now we have either n = 15-n or n = 13-n
and since n is an integer, these are both impossible. In Figure 19(ii), the diagonal stripes must obviously collide or, more formally, we must have n = 13-n or
n = 9-n, again impossible. This leaves only case (a) to consider, and again two
cases are possible as in Figure 20.

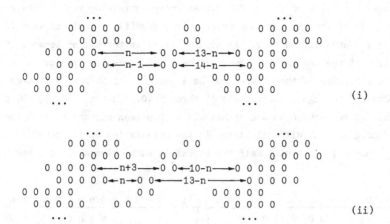

Figure 20: configurations arising in case (9a).

In Figure 20(i), the blocks of 2-strings are aligned with 5-strings which
overlap each other by four columns, and in Figure 20(ii), with 5-strings which
overlap each other by two columns. For sequentiality in (i), we have only two
distinct values for n, n-1, 13-n amd 14-n, forcing n = 7; in (ii) similarly, we
must have n = 5. However n = 7 in the rows of the array forces n = 5 in the
columns and vice versa, so sequentiality is not possible. Hence case (9) cannot
occur.

Case (5a): The string above the 5-string must be completed, giving Figure 22(i);
now the 2-string below the 5-string can be put into the required configuration in
one of two ways, leading to either Figure 22(ii) or (iii). In either case, the
array must in fact be doubly uniform and each block of 0s can have two orientations,
as shown in Figure 23.

```
                                 1                       1
1 0 0 0 0 0 1                  1 0 1                   1 0 1
  1 0 0 0 0 0 1               1 1 0 0 1               1 1 0 0 1
    1 0 0 1 1             1 0 0 0 0 0 1             1 0 0 0 0 0 1
      (i)                  1 0 0 0 0 0 1             1 0 0 0 0 0 1
                             1 0 0 1 1                 1 0 0 1 1
                             1 0 1                     1 0 1
                             1      (ii)               1      (iii)
```

Figure 22: blocks arising in case (5a).

```
   0                 0                  0                  0
  0 0               0 0                0 0                0 0
0 0 0 0 0         0 0 0 0 0          0 0 0 0 0          0 0 0 0 0
  0 0 0 0 0         0 0 0 0 0          0 0 0 0 0          0 0 0 0 0
    0 0               0 0                0 0                0 0
     0                 0                  0                  0

    (i)               (i')               (ii)              (ii')
```

Figure 23: orientations of blocks from Figure 22.

Consider the block of Figure 23(i). An array containing such blocks must either have all of them in the same orientation, say (i), or half of them in orientation (i) and half in (i'). An argument similar to that of Lemma 2, Case (7a), shows that, whatever the orientations, if the first block lies in rows 1 through 6 and columns 1 through 6, then there must be a block lying in rows 3 through 8 and either columns 7 through 12 or columns 15 through 20. In other words, the block must be shifted two rows down and either six or fourteen columns right. The possible configurations with all blocks in the orientation of Figure 23(i) are shown in Figure 24; those with half the blocks in each orientation are shown in Figure 25.

```
1 2 3 4 5 6 7 8 9 0 1 2 3 4 5 6 7 8 9 0       1 2 3 4 5 6 7 8 9 0 1 2 3 4 5 6 7 8 9 0
    0              0 0      0 0 0 0    |1|         0          0 0 0 0 0        0 0
    0 0              0        0 0 0 0 0 |2|         0 0          0 0 0 0 0        0
0 0 0 0 0        0              0 0    |3|  0 0 0 0 0          0 0              0
  0 0 0 0 0      0 0            0      |4|    0 0 0 0 0        0              0 0
    0 0      0 0 0 0 0      0          |5|      0 0          0          0 0 0 0 0
      0        0 0 0 0 0    0 0        |6|        0            0 0          0 0 0 0 0
0              0 0      0 0 0 0 0      |7|          0          0 0 0 0 0        0 0
0 0              0        0 0 0 0 0    |8|          0 0          0 0 0 0 0        0
0 0 0      0            0 0      0 0   |9|      0 0 0 0 0          0 0              0
0 0 0 0    0 0            0        0   |0|        0 0 0 0 0        0              0 0
0 0    0 0 0 0 0      0                |1|  0          0 0          0        0 0 0 0
  0        0 0 0 0 0    0 0            |2|  0 0          0          0 0        0 0 0
          0 0    0 0 0 0 0      0      |3|            0          0 0 0 0 0        0 0
            0      0 0 0 0 0    0 0    |4|            0 0          0 0 0 0 0        0
0      0              0 0      0 0 0 0 |5|  0          0 0 0 0 0          0 0
0 0    0 0              0        0 0 0 |6|  0 0          0 0 0 0 0          0
  0 0 0 0 0      0            0 0      |7|  0 0 0          0 0              0        0 0
    0 0 0 0 0    0 0            0      |8|  0 0 0 0          0              0 0        0
      0 0    0 0 0 0 0      0          |9|  0 0            0          0 0 0 0 0
        0      0 0 0 0      0 0        |0|  0            0 0          0 0 0 0 0
```

(i) Block shifted by two rows down   (ii) Block shifted by two rows down
     and six columns right.   and fourteen columns right.

Figure 24: configurations with all blocks in orientation of Figure 23(i).

The row sequence of Figure 24(i) must be

0 0 0 0 0 - - - 0 - - - - - - - 0 0 - -, but the column sequence is

0 0 0 0 0 - - - 0 0 - - - - - - 0 - - - . Hence the sequences cannot agree and the array is not sequential. In Figure 24(ii), the row and column sequences are interchanged. In any case, no such array is possible.

The row sequences of Figure 25(i) are

0 0 0 0 0 - - - 0 0 - - - - - - 0 - - and 0 0 0 0 0 - - 0 0 - - - - - - 0 - - - -

alternately, and so are the column sequences. In Figure 25(ii), we have the same sequences, though the array is different. Neither of these arrays can be sequential.

```
1 2 3 4 5 6 7 8 9 0 1 2 3 4 5 6 7 8 9 0        1 2 3 4 5 6 7 8 9 0 1 2 3 4 5 6 7 8 9 0
    0                 0 0         0 0 0 0 0 |1|     0               0 0 0 0 0       0 0
    0 0                 0         0 0 0 0 0 |2|     0 0         0 0 0 0 0             0
0 0 0 0 0                 0             0 0 |3| 0 0 0 0 0             0 0                   0
  0 0 0 0 0       0 0                   0   |4| 0 0 0 0 0         0                     0 0
    0 0         0 0 0 0 0       0           |5|     0 0                 0         0 0 0 0 0
      0       0 0 0 0 0             0 0     |6|     0               0 0         0 0 0 0 0
  0             0 0       0 0 0 0 0         |7|         0         0 0 0 0 0           0 0
0 0               0         0 0 0 0 0       |8|         0 0         0 0 0 0 0         0
0 0 0 0       0             0 0         0   |9|         0 0 0 0 0         0 0               0
0 0 0         0 0             0       0 0   |0|         0 0 0 0 0             0           0 0
0 0       0 0 0 0 0             0           |1| 0         0 0                 0         0 0 0 0
0         0 0 0 0 0         0 0             |2| 0 0         0             0 0           0 0 0
          0 0         0 0 0 0 0       0     |3|             0             0 0 0 0 0       0 0
            0       0 0 0 0 0         0 0   |4|             0 0         0 0 0 0 0             0
0               0             0 0       0 0 0 0 |5|     0         0 0 0 0 0         0 0
0 0         0 0             0           0 0 0 |6| 0 0         0 0 0 0 0         0
    0 0 0 0 0         0                 0 0 |7| 0 0 0 0         0 0             0               0
  0 0 0 0 0         0 0                   0 |8| 0 0 0             0           0 0           0 0
      0 0         0 0 0 0 0         0       |9| 0 0           0         0 0 0 0 0
      0         0 0 0 0 0         0 0       |0| 0             0 0         0 0 0 0 0
```

(i) Block shifted by two rows down   (ii) Block shifted by two rows down
     and six columns right.   and fourteen columns right.

Figure 25: configurations with half the blocks in orientation of Figure 23(i),
            half in that of Figure 23(i')

Finally, consider the block of Figure 23(ii). As before, all the blocks in the
array may have the same orientation, say (ii), or half of them may have orientation
(ii) and half (ii'). In either case, successive blocks in the array must be shifted
two rows down and either six or fourteen columns right. The possible configurations
are shown in Figures 26 and 27. In Figure 26(i), each row and column has the
sequence 0 0 0 0 0 - - 0 0 - - - - - - 0 - - - - , but the array cannot be
completed, since we cannot fit a 5- and a 2-string of 1s into this sequence. In
Figure 26(ii), each row and column has the sequence
0 0 0 0 0 - - - 0 0 - - - - - - - 0 - - . The 2-string of 1s must fit between the
isolated 0 and the 5-string of 0s. Also note that in row 3 of the array, columns
11 and 12 must contain 1s, so the 5-string must immediately follow the 2-string of
0s, giving the sequence 0 0 0 0 0 1 0 1 0 0 1 1 1 1 1 0 1 0 1 1 and the array of
Figure 28, as in the statement of the lemma.

In Figure 27(i), we have again two different sequences occurring in the rows,
namely 0 0 0 0 0 - - - 0 0 - - - - - - 0 - - - and
0 0 0 0 0 - - 0 0 - - - - - - - 0 - - - . The same sequences occur in the columns
and clearly the array is not sequential. Again in Figure 27(ii), though the array
is different, the same two sequences occur.

This completes the proof of the Lemma.

```
1 2 3 4 5 6 7 8 9 0 1 2 3 4 5 6 7 8 9 0        1 2 3 4 5 6 7 8 9 0 1 2 3 4 5 6 7 8 9 0
      0               0 0         0 0 0 0 0  1       0         0 0 0 0 0         0 0
    0 0             0             0 0 0 0 0  2     0 0         0 0 0 0 0       0
0 0 0 0 0               0             0 0    3  0 0 0 0 0           0 0                   0
  0 0 0 0 0       0 0                  0     4     0 0 0 0 0       0               0 0
    0 0       0 0 0 0 0               0      5       0 0               0       0 0 0 0 0
    0         0 0 0 0 0         0 0          6       0                 0 0       0 0 0 0 0
  0               0 0       0 0 0 0 0        7           0       0 0 0 0 0           0 0
0 0               0         0 0 0 0 0        8         0 0       0 0 0 0 0       0
0 0 0             0             0 0       0 0 9     0 0 0 0 0       0 0                   0
0 0 0 0       0 0                 0       0   0     0 0 0 0 0       0                 0 0
0 0       0 0 0 0 0             0         0 0 1  0       0 0               0       0 0 0 0
0         0 0 0 0 0         0 0           0 0 2  0 0       0               0 0       0 0 0
            0 0       0 0 0 0 0           0  3         0       0 0 0 0 0           0 0
            0         0 0 0 0 0       0 0    4         0 0       0 0 0 0 0         0
0             0             0 0       0 0 0 0 5     0       0 0 0 0 0       0 0
0 0       0 0               0             0 0 0 6  0 0       0 0 0 0 0       0
    0 0 0 0 0             0             0 0  7  0 0 0       0 0               0       0 0
    0 0 0 0 0       0 0                 0    8  0 0 0 0       0               0 0       0
      0 0       0 0 0 0 0             0      9  0 0       0                 0 0       0 0 0 0 0
      0         0 0 0 0 0         0 0        0  0               0 0       0 0 0 0 0
```

(i) Block shifted by two rows down   (ii) Block shifted by two rows down
     and six columns right.   and fourteen columns right.

Figure 26: configurations with all blocks in orientation of Figure 23(ii).

```
1 2 3 4 5 6 7 8 9 0 1 2 3 4 5 6 7 8 9 0        1 2 3 4 5 6 7 8 9 0 1 2 3 4 5 6 7 8 9 0
      0               0 0         0 0 0 0 0  1       0         0 0 0 0 0         0 0
    0 0             0             0 0 0 0 0  2     0 0         0 0 0 0 0       0
0 0 0 0 0               0             0 0    3  0 0 0 0 0           0 0                   0
  0 0 0 0 0       0 0                  0     4     0 0 0 0 0       0               0 0
    0 0       0 0 0 0 0               0      5       0 0               0       0 0 0 0 0
    0         0 0 0 0 0         0 0          6       0                 0 0       0 0 0 0 0
0                 0 0       0 0 0 0 0        7           0       0 0 0 0 0           0 0
0 0               0         0 0 0 0 0        8         0 0       0 0 0 0 0       0
0 0 0 0           0             0 0       0  9     0 0 0 0 0       0 0                   0
0 0 0             0 0                 0   0 0 0     0 0 0 0 0       0                 0 0
0 0 0       0 0 0 0 0             0         0 1  0       0 0               0       0 0 0 0
0         0 0 0 0 0         0 0           0 0 2  0 0       0               0 0       0 0 0
            0 0       0 0 0 0 0           0  3         0       0 0 0 0 0           0 0
            0         0 0 0 0 0       0 0    4         0 0       0 0 0 0 0         0
0             0             0 0       0 0 0 0 5  0       0 0 0 0 0       0 0
0 0       0 0               0             0 0 0 6  0 0       0 0 0 0 0       0
    0 0 0 0 0             0             0 0  7  0 0 0 0       0 0               0       0 0
    0 0 0 0 0       0 0                 0    8  0 0 0       0               0 0       0 0
      0 0       0 0 0 0 0             0      9  0 0       0                 0 0       0 0 0 0 0
      0         0 0 0 0 0         0 0        0  0               0 0       0 0 0 0 0
```

(i) Block shifted by two rows down   (ii) Block shifted by two rows down
     and six columns right.   and fourteen columns right.

Figure 27: configurations with half the blocks in orientation of Figure 23(ii),
            half in that of Figure 23(ii').

Figure 28: doubly uniform sequential balanced array of period 20, unique up to
reflection, with sequence 0 0 0 0 0 1 0 1 0 0 1 1 1 1 1 0 1 0 1 1.

We summarise the results of this section in the following

THEOREM. *There exists exactly one row-uniform and column-uniform balanced array of period 20, namely that of Figure 28.*

ARRAYS CONSIDERED AS TILINGS BY POLYOMINOES.

A *polyomino* is a plane figure obtained by joining unit squares along their edges in rook-wise fashion, that is, so that a chess rook could travel through all the squares in the piece. A piece with a hole in it is not considered to be a polyomino. For an account of packing problems and polyominoes, see for example Honsberger [4], Chapter 8.

In Figure 28 the blocks of 0s (and similarly 1s) form 16-ominoes which, together with the isolated single squares, tile the plane. These 16-ominoes have the curious property shown in Figure 29, namely, they can themselves be tiled with four tetrominoes in three ways using L-, T- or skew-tetrominoes respectively.

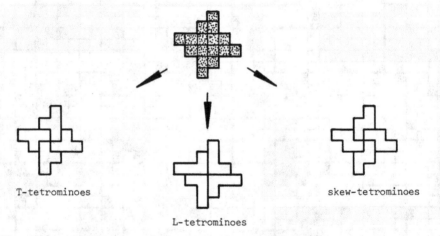

T-tetrominoes   skew-tetrominoes

L-tetrominoes

Figure 29:   tiling the 16-omino of Figure 28 with tetrominoes.

In Figure 30 we show one of the uniform tilings of the plane which is not made up of edge-to-edge tiles. (**Eight** such families exist; see for example the expository article [3].) The heavy lines covering Figure 28 show the very close relationship between this array and the uniform tiling.

Subsequent diagrams show, in the form of tilings, binary arrays which are of interest for various reasons, although they do not have all the properties that we are looking for. Each of the tilings in Figures 31 and 32 is balanced for one symbol but not the other. Figure 31 arises from the configuration of Figure 20 and Figures 32(i) and (ii) from that of Figure 14(iv), when we attempt to fill in the sequences. Because each of them has period 20, we have very little room to manoeuvre and cannot achieve balance in both symbols.

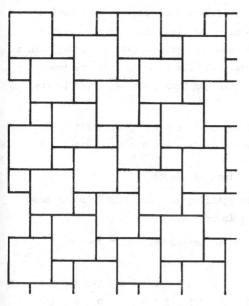

Figure 30:   uniform tiling related to
the array of Figure 28.

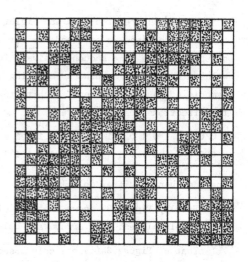

Figure 31:   array with row sequence
1 1 1 1 1 0 0 1 1 0 1 0 1 0 0 0 0 0 1 0
and column sequence
1 1 1 1 1 0 0 0 0 0 1 0 1 0 1 0 1 1 0 0,
arising from Figure 20 and
balanced for 0 but not for 1.

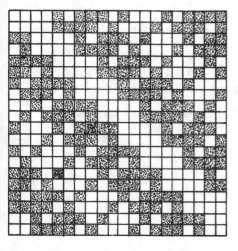

(i)   array with row sequence
0 0 0 0 1 0 1 1 1 0 1 0 0 0 1 0 1 1 1 1
and column sequence
0 0 0 0 1 0 1 1 1 0 0 0 1 0 1 0 1 1 1 1

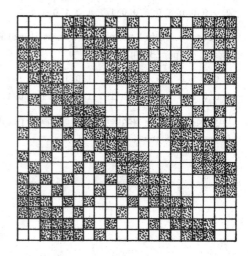

(ii)   array with row sequence
0 0 0 0 1 1 1 0 1 0 1 0 0 0 1 0 1 1 1 1
and column sequence
0 0 0 0 1 1 1 0 1 0 0 0 0 1 0 1 0 1 1 1

Figure 32:   arrays arising from Figure 14(iv), balanced for 0 but not for 1.

The closely related Pooh-and-Piglet array of Figure 33 however, with period 40 in each direction, is balanced for both symbols. Parallel lines of the array are occupied by the same sequence, always in the same direction. (All the other arrays we have found have sequences reversing direction.) The 16-ominoes of the two symbols are quite different, despite the balance for both. The sequences in the two directions are respectively

0 0 0 0 0 1 0 1 0 1 1 0 0 0 0 1 1 1 1 1 0 0 0 1 0 1 1 1 0 1 0 1 0 0 1 1 1 1 0 1 and
0 0 0 0 0 1 0 1 0 1 1 0 0 1 0 1 1 1 1 1 0 1 0 0 0 1 1 1 0 1 0 0 0 0 1 1 1 1 0 1;
               ↑   ↑ ↑   ↑

these sequences differ only in the four positions indicated.

Finally Figure 34 shows that two distinct tilings can be built from the same set of polyominoes; both arrays are balanced in both symbols.

That of Figure 34(i) has column sequences of period 40, namely the complementary sequences

0 0 0 0 0 1 1 1 1 0 1 1 1 1 1 0 0 0 0 1 0 1 0 0 0 1 0 1 1 0 1 0 1 1 1 0 1 0 0 1 and
1 1 1 1 1 0 0 0 0 1 0 0 0 0 0 1 1 1 1 0 1 0 1 1 1 0 1 0 0 1 0 1 0 0 0 1 0 1 1 0;

its row sequences are of period 20, namely

0 0 0 0 0 1 1 1 1 0 1 0 1 1 1 0 1 0 0 1 , its complementary sequence
1 1 1 1 1 0 0 0 0 1 0 1 0 0 0 1 0 1 1 0 ,
1 1 1 1 1 0 1 1 1 1 0 0 0 1 0 1 0 0 1 0 and its complementary sequence
0 0 0 0 0 1 0 0 0 0 1 1 1 0 1 0 1 1 0 1 .

In Figure 34(ii) where the same polyominoes are rearranged, the column sequences, again of period 40, are

0 0 0 0 0 1 0 0 0 0 1 1 1 0 1 0 1 1 1 1 0 0 0 1 0 1 0 0 1 0 1 1 1 1 1 0 1 1 0 1 and
0 0 0 0 0 1 0 0 0 0 1 1 1 1 1 0 1 1 0 1 0 0 0 1 0 1 0 0 1 0 1 1 1 0 1 0 1 1 1 1 ;

but now the row sequences also have period 40 since they are the complements of the column sequences.

Figure 33: Pooh-and-Piglet array of period 40, balanced for both symbols.

Figure 34(i): balanced array from two polyominoes, period 20 along rows, 40 along columns.

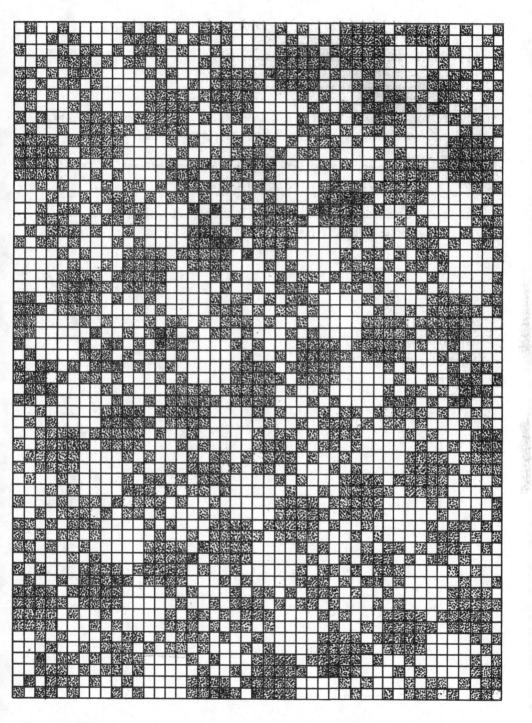

Figure 34(ii): balanced array from two polyominoes, period 40 in each direction.

REFERENCES

[1] R.M. Cormack, private communication, 1978.

[2] D.J. Gates, private communication, 1978.

[3] Branko Grünbaum and Geoffrey C. Shephard, Tilings by regular polygons,
       Math. Mag. 50 (1977) 227-247; 51 (1978) 205-206.

[4] Ross Honsberger, *Mathematical Gems II,* (Dolciani Mathematical Expositions
       Volume 2, Math. Assoc. of America, 1976).

[5] Joan P. Hutchinson, On words with prescribed overlapping subsequences,
       Utilitas Math. 7 (1975) 241-250.

Department of Mathematics
University of Queensland
St. Lucia
Queensland

CLASSIFYING K-CONNECTED CUBIC GRAPHS

NICHOLAS C. WORMALD

This paper examines simple operations by which the cubic graphs with a given connectedness on p points can be constructed from those on less than p points. In particular, for cyclically 4-connected cubic graphs, the only operation required is that of joining the midpoints of two non-adjacent lines.

1. INTRODUCTION

A graph G is *k-connected* if it has at least k+1 points and the removal of k-1 points and their incident lines from G always leaves a connected graph. Also, G is *cubic* if all its points have degree 3. The problem of constructing the k-connected cubic graphs on p points from those on less than p points is considered in Section 3 for k = 1 and 2. A simple solution has already been given by Tutte [6, p.140] when k is 3. Let x and y be two distinct lines of a graph G. By *joining* x and y, we mean that process which consists of subdividing x and y by points u and v respectively, and then joining u and v. Tutte's result can be expressed as follows.

Theorem 1. *Each 3-connected cubic graph G other than K_4 can be obtained from a 3-connected cubic graph H by joining two lines of H, and each graph G obtainable in this way is 3-connected.*

The corresponding results given in Section 3 for 1- and 2-connected cubic graphs are more complicated, in that they require additional operations.

Clearly no cubic graph is 4-connected. However, the following measure of connectivity is non-trivial for cubic graphs when k is large. First, an *n-cutset* of a graph G is a collection S of n lines of G whose removal results in a disconnected graph H. Each component of H is an *n-end* of G (at S). We say S is *cycle-separating* if at least two of the n-ends of G at S contain cycles, and G is *cyclically k-connected* if it has no cycle-separating n-cutset for $0 \leq n < k$. Cyclically 4- and 5-connected graphs have been studied in connection with the four-colour map problem (see Grünbaum [2, p.365]). Of special interest to some physicists (see Brink and Satchler [1, Chapter 7]) are the cyclically 4-connected cubic graphs, which have been termed *irreducible* graphs. In Section 4 we extend the method of Section 3 to show that whenever two non-adjacent lines of an irreducible cubic graph are joined, the resulting graph is irreducible, and that with just two exceptions, each irreducible cubic graph can be obtained in this way from a smaller irreducible cubic graph. The two exceptional graphs are K_4, the complete graph on four points, and Q_3, the 1-skeleton of the cube. We have not been able to find a similar characterisation of cubic graphs

of higher connectivities; this and other related problems are discussed in Section 5.
Graph theoretic notation not defined here can be found in the book by Harary [3].

2.   PRELIMINARY RESULTS

The results of this section will be used in each of the two following sections.

Lemma 1. *Suppose a cyclically k-connected cubic graph G cannot be obtained
from another such graph H by joining two lines of H. Let x be a line of G such that
neither endpoint of x lies in a triangle not containing x, and x does not lie in two
triangles of G. Then x is a member of a cycle-separating k-cutset of G.*

Proof. Suppose the hypotheses of the lemma are satisfied for some line
x = uv of G. Then x can be removed and u and v suppressed into two lines, y and z
respectively, and no multiple line is formed in the resulting cubic graph H. This
process is the reverse of joining the lines y and z of H, so H cannot be cyclically
k-connected. Thus, for some n < k, H has a cycle-separating n-cutset S. But if
neither y nor z is in S, this implies that $T = S \cup \{x\}$ is a cycle-separating (n+1)-
cutset of G. Hence n+1 = k and the lemma follows in this case. On the other hand,
if y is in S the same conclusion is reached by taking

$$T = (S - \{y\}) \cup \{x,w\},$$

where w is a line of G incident with u. If z is in S, it can be replaced in the same
way by a line incident with v.

The next result will allow us to use Lemma 1 in connection with k-connected
cubic graphs. Its proof is trivial.

Lemma 2. *For k = 1, 2 and 3, every (k-1)-cutset of a cubic graph is cycle-
separating, so a cubic graph is k-connected if and only if it is cyclically k-connected.*

Lemma 3. *Each cycle-separating k-cutset S of a cyclically k-connected cubic
graph G is independent, and the removal of the lines in S from G leaves just two k-
ends.*

Proof. Suppose a cycle-separating k-cutset S of a cubic graph G includes
two adjacent lines, say x = ut and y = vt. Then either x or y can be omitted from S,
or both can be replaced in S by the third line adjacent to t, to form a cycle-separating
(k-1)-cutset of G.

Suppose the removal of the lines in S from G leaves at least three k-ends.
If one of these, say H, is acyclic, then one of the lines of S adjacent to H can be
omitted from S to form a cycle-separating (k-1)-cutset of G. If none of the k-ends
is acyclic, any line can be omitted from S.

3.   CLASSIFYING k-CONNECTED CUBIC GRAPHS

By the *addition of K_4* to a line x of a graph G, we mean the process of taking the graph $G \cup K_4$ and joining x to a line of K_4. By the *insertion of two triangles* in a line x = uv of G, we mean the process of replacing x by the graph shown in Figure 1, the points of this graph other than u and v being distinct from those of G.

Figure 1.   Insertion of two triangles.

When either of these operations is applied to a connected cubic graph H, or if two lines of H are joined, a connected cubic graph results. Hence, the following result is a complete characterisation of connected cubic graphs.

Theorem 2. Each connected cubic graph other than K_4 can be obtained from a connected cubic graph with fewer points by the addition of K_4 to a line, or by the insertion of two triangles in a line, or by joining two lines.

Proof.   Let G be a connected cubic graph not obtainable by applying one of these operations to any other such graph, and suppose G is not K_4.

Assume that some line x lies in two triangles of G. Since G is not obtainable from another connected cubic graph by the insertion of two triangles or the addition of K_4, it is seen that the structure of G near x must be as shown in Figure 2(a) or (b). But in each of these cases, G can be formed by joining two lines of a connected cubic graph using the line y, contrary to hypothesis. Thus no line of G lies in two triangles.

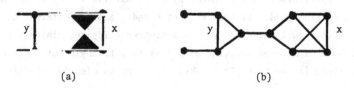

| (a) | (b) |

Figure 2.   Two impossible situations.

Now suppose a line x of G is not a bridge (1-cutset) of G. Then by Lemmas 1 and 2, at least one of the endpoints of x must lie in a triangle not containing x. It is clear that K_4 is the only cubic graph in which all lines have the latter property. Consequently, G must have at least one bridge, and hence at least two 1-ends. Let H be a 1-end of G with the minimal number of lines. Then none of the lines of H can be a bridge of G; otherwise there would be an even smaller 1-end of G within H. Therefore, for each line x in H, some endpoint of x lies in a triangle of G which does not contain x, and x is not contained in two triangles. It is readily verified that this situation is impossible. The proof is complete.

To obtain a similar characterisation of 2-connected cubic graphs, one first needs to observe that the joining of two lines, or the insertion of two triangles in a line, of a 2-connected cubic graph produces another 2-connected cubic graph. The characterisation is completed as follows.

Theorem 3. *Each 2-connected cubic graph other than K_4 can be obtained from a 2-connected cubic graph with fewer points by inserting two triangles in a line or by joining two lines.*

Proof. This is similar to the proof of Theorem 2. Let G be a 2-connected cubic graph not obtainable by applying either of these two operations to any other such graph, and suppose G is not K_4.

Assume that some line x lies in two triangles of G. Since G is 2-connected but is not obtainable from another 2-connected cubic graph by the insertion of two triangles, it is seen that the structure of G near x must be as shown in Figure 2(a). But then G can be formed by joining two lines of a 2-connected graph using the line y. Hence no line of G lies in two triangles. Therefore, by Lemmas 1 and 2, for each line x of G, either x is in a 2-cutset of G or some endpoint of x lies in a triangle not containing x. As in the proof of Theorem 2, all lines cannot be in the latter category, and we deduce that G has a 2-cutset and consequently some 2-ends.

Let H be a 2-end of G at a 2-cutset $S = \{x,y\}$ and suppose the number of lines of H is minimal for a 2-end of G. By Lemma 3, two distinct points, say u and v, are incident with x and y respectively, and there is one 2-end, say H*, of G at S other than H. Suppose a line z of H and some other line z* form a 2-cutset T of G. Then T is cycle-separating by Lemma 2, so z must lie in H*, or the minimality of H would be contradicted. It follows that u and v are in different components of H-z, and so $\{x,z\}$ is a 2-cutset of G, again contradicting the minimality of H. Thus for every line z of H, some endpoint of z lies in a triangle not containing z, and z is not contained in two triangles. This situation is clearly impossible, and the proof is complete.

4. CLASSIFYING IRREDUCIBLE CUBIC GRAPHS

Two more preliminary results are covered before the characterisation of irreducible cubic graphs is begun.

Lemma 4. *Every 3-cutset of an irreducible cubic graph consists of the three lines incident with some point.*

Proof. If G is a cubic graph with a cutset S, then any acyclic component of G-S with more than one point must be adjacent to at least four lines of S.

Lemma 5. *Let H be a 4-end of an irreducible cubic graph G at an independent 4-cutset S. Then H has no bridge, and no independent 2-cutset of H can separate one of the points of degree 2 in H from the other three.*

Proof. Denote the points of degree 2 in H by $u_1, \ldots, u_4$, and the line of S incident with u_i by x_i. As G has no bridge, any bridge y of H must separate one or two of the u's from the others. If u_1 is separated from the others, then y and x_1 form a 2-cutset of G, which is a contradiction by Lemma 2. If u_1 and u_2 are separated from u_3 and u_4 by y, then x_1, x_2 and y form a 3-cutset of G, and this is a contradiction by Lemma 4. Hence H can have no bridge. If an independent 2-cutset S of H separates u_1 from u_2, u_3 and u_4, then $S \cup \{x_1\}$ is a 3-cutset of G and again we reach a contradiction by Lemma 4.

Our characterisation of irreducible cubic graphs is based on the operation of joining non-adjacent lines. Suppose two non-adjacent lines uu' and vv' of an irreducible cubic graph G are joined by a line x = u*v* to form a graph H, so that u* is adjacent to u and u' in H. Assume that H has a cycle-separating 3-cutset S. By Theorem 1, H is 3-connected, and so by Lemmas 2 and 3, S is independent. Hence S must contain at least one of the lines x, uu*, u'u*, vv* and v'v*, for S would otherwise be a cycle-separating 3-cutset of G. If x is in S, then S-{x} is a 2-cutset of G, which is a contradiction. If uu*, say, is in S, and not vv* or v'v*, then (S-{uu*}) ∪ {uu'} is a 3-cutset of G. This is a contradiction by Lemma 4. A similar contradiction is reached if S contains, say, uu* and vv* or v'v*, since these lines can be replaced in S by uu' and vv' to form a 3-cutset of G.

Thus, it is seen that joining two non-adjacent lines of an irreducible cubic graph always produces another irreducible cubic graph. Our characterisation of these graphs is completed as follows.

Theorem 4. *Each irreducible cubic graph other than K_4 and Q_3 can be obtained from an irreducible cubic graph with fewer points by joining two non-adjacent lines.*

Proof. Let G be an irreducible cubic graph not obtainable by this operation, and suppose G is not K_4. We aim to show that G is isomorphic to Q_3.

Note that G can contain no triangles, so G is not even obtainable by joining two adjacent lines of an irreducible graph. Hence, by Lemma 1, each line of G is a member of a cycle-separating 4-cutset. Furthermore, each such 4-cutset is independent by Lemma 3.

Let H be a 4-end of G at a cycle-separating 4-cutset S = $\{x_1, \ldots, x_4\}$, and suppose the number of lines of H is minimal for a cycle-containing 4-end of G. Denote the point of H incident with x_i by u_i for each i. By Lemma 3, there is one 4-end, say H*, of G at S other than H. Let y_1 be a line of H. Then from the preceding paragraph,

204

there is some independent 4-cutset $T = \{y_1,\ldots,y_4\}$ of G.

If there is no line of H in T other than y_1, then $T-\{y_1\}$ is a cutset of G by Lemma 5, which is a contradiction by Lemma 4. Similarly, there cannot be precisely one line of H* in T. If no lines of H* are in T, the minimality of H is contradicted. Hence, just two lines of T are in H and two are in H*. Say y_1 and y_2 are in H. These lines must separate the u's in H somehow, so by Lemma 5 we can assume that y_1 and y_2 separate u_1 and u_2 from u_3 and u_4 in H. Thus $\{x_1,x_2,y_1,y_2\}$ is a 4-cutset of G. As H is minimal, this 4-cutset cannot be cycle-separating, and thus it is seen that x_1 is adjacent to y_1 or y_2 and x_2 is adjacent to the other. This is for any line y_1 of H, so we conclude that each line of H is adjacent to some line of S. It follows that H is just a 4-cycle consisting of the lines y_1,y_2,u_1u_2 and u_3u_4. Denote the two latter lines by z_1 and z_2, and denote by u_i* the point of H* incident with x_i. Then one component of H*-$\{y_3,y_4\}$, say F_1, must contain u_1* and u_2*, and the other, say F_2, contains u_3* and u_4*. This situation is illustrated in Figure 3.

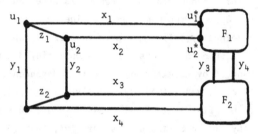

Figure 3. A complicated way to draw Q_3.

Now, z_1 must be in some independent 4-cutset T' of G which necessarily includes z_2. T' therefore separates u_1* from u_2* in G, and u_3* from u_4*. As F_1 and F_2 are connected, it follows that one line of T', say z_3, is in F_1, and the other, say z_4, is in F_2. Hence z_3 is a bridge of F_1. But $R = \{x_1,x_2,y_3,y_4\}$ is a 4-cutset of G, so by Lemma 5, R is not independent, and thus R is not cycle-separating by Lemma 3. It follows that F_1 consists of just u_1* and u_2* together with the line $z_3 = u_1$*u_2*. A similar argument suffices to show that z_4 is the only line in F_2. The fact that T' is a 4-cutset of G now forces G to be isomorphic to Q_3. This establishes the theorem.

5. RELATED RESULTS

We define the *distance* between two lines x and y of a connected graph to be one less than the length of the shortest path which includes x and y. Two non-adjacent lines have distance at least 2, and it is readily shown that in general, joining two lines with distance at least k-2 in a cyclically k-connected cubic graph produces another such graph. For $k \geqslant 3$, let S_k denote the set of cyclically k-connected cubic graphs not obtainable by this process. By Theorem 1, $S_3 = \{K_4\}$, and Theorem 4 states that $S_4 = \{K_4, Q_3\}$. So far, this would appear to be a promising approach to the investigation of cyclically k-connected graphs for k in general. For $k \geqslant 5$, however, we have not determined S_k precisely, but we have found an infinite subset T of S_5. The set T is of a form which leads the author to believe that S_5 is difficult to characterise.

The graphs in T can be constructed as follows. Let G be any 5-connected 5-regular graph with p points, and let H be the graph shown in Figure 4.

Figure 4. The 1-skeleton of a dodecahedron, with a 5-cycle removed.

A cubic graph F can now be constructed by replacing each point of G with a different copy of H, and replacing each line of G with a line joining two points of degree 2 in the corresponding copies of H. It can always be arranged that F is cubic, and in general this can be done in many ways. It is not difficult to show that F is cyclically 5-connected. Furthermore, at least one endpoint of each line of F lies in a 5-cycle not containing that line, so F cannot be formed by joining two lines of a cyclically 5-connected cubic graph. Hence F is in S_5. The existence of G may be in doubt for a given p, but some of the results of [8] imply that when p is large and even the number of different possibilities for G is at least $cp^{3p/2}$, where c is a positive constant. The above construction does not readily generalise for S_k (we have run out of suitable Platonic solids), so the question of the size of S_k for $k \geqslant 6$ is open (and should perhaps remain so, considering the size of S_5).

Most of the problems we have considered in this paper have already been examined for k-connected graphs in general. For some values of k there have been a number of different classifications of the k-connected graphs. In particular, Tutte [6] has given two separate characterisations of 3-connected graphs (both distinct from that of the author [7]), one of which implies Theorem 1, and Slater [4 and 5] has considered the problem for other values of k.

Elsewhere we shall enumerate labelled irreducible cubic graphs. This will be accomplished by using a characterisation of these graphs which is different from that presented in Section 4.

REFERENCES

[1] G.M. Brink and D.R. Satchler, *Angular Momentum*. 2nd edn (Oxford University Press, London, 1968.)

[2] B. Grünbaum, *Convex Polytopes*. (John Wiley & Sons, London, 1967.)

[3] F. Harary, *Graph Theory*. (Addison-Wesley, Reading, Mass., 1969.)

[4] P.J. Slater, A classification of 4-connected graphs, *J. Combinatorial Theory* 17B (1974), 281-298.

[5] P.J. Slater, General soldering, *Proceedings of the Fifth British Combinatorial Conference* (C.St.J.A. Nash-Williams and J. Sheehan, eds) Utilitas Math., Winnipeg; 1976, 559-567.

[6] W.T. Tutte, *Connectivity in Graphs*. (Toronto University Press, Toronto, 1967.)

[7] N. Wormald, Enumeration of labelled graphs I: 3-connected graphs, *J. London Math. Soc.* (to appear).

[8] N.C. Wormald, The asymptotic connectivity of labelled regular graphs, (to appear).

Department of Mathematics,
University of Newcastle,
New South Wales.

Vol. 580: C. Castaing and M. Valadier, Convex Analysis and Measurable Multifunctions. VIII, 278 pages. 1977.

Vol. 581: Séminaire de Probabilités XI, Université de Strasbourg. Proceedings 1975/1976. Edité par C. Dellacherie, P. A. Meyer et M. Weil. VI, 574 pages. 1977.

Vol. 582: J. M. G. Fell, Induced Representations and Banach *-Algebraic Bundles. IV, 349 pages. 1977.

Vol. 583: W. Hirsch, C. C. Pugh and M. Shub, Invariant Manifolds. V, 149 pages. 1977.

Vol. 584: C. Brezinski, Accélération de la Convergence en Analyse Numérique. IV, 313 pages. 1977.

Vol. 585: T. A. Springer, Invariant Theory. VI, 112 pages. 1977.

Vol. 586: Séminaire d'Algèbre Paul Dubreil, Paris 1975-1976 (29ème Année). Edited by M. P. Malliavin. VI, 188 pages. 1977.

Vol. 587: Non-Commutative Harmonic Analysis. Proceedings 1976. Edited by J. Carmona and M. Vergne. IV, 240 pages. 1977.

Vol. 588: P. Molino, Théorie des G-Structures: Le Problème d'Equivalence. VI, 163 pages. 1977.

Vol. 589: Cohomologie l-adique et Fonctions L. Séminaire de Géométrie Algébrique du Bois-Marie 1965-66, SGA 5. Edité par L. Illusie. XII, 484 pages. 1977.

Vol. 590: H. Matsumoto, Analyse Harmonique dans les Systèmes de Tits Bornologiques de Type Affine. IV, 219 pages. 1977.

Vol. 591: G. A. Anderson, Surgery with Coefficients. VIII, 157 pages. 1977.

Vol. 592: D. Voigt, Induzierte Darstellungen in der Theorie der endlichen, algebraischen Gruppen. V, 413 Seiten. 1977.

Vol. 593: K. Barbey and H. König, Abstract Analytic Function Theory and Hardy Algebras. VIII, 260 pages. 1977.

Vol. 594: Singular Perturbations and Boundary Layer Theory, Lyon 1976. Edited by C. M. Brauner, B. Gay, and J. Mathieu. VIII, 539 pages. 1977.

Vol. 595: W. Hazod, Stetige Faltungshalbgruppen von Wahrscheinlichkeitsmaßen und erzeugende Distributionen. XIII, 157 Seiten. 1977.

Vol. 596: K. Deimling, Ordinary Differential Equations in Banach Spaces. VI, 137 pages. 1977.

Vol. 597: Geometry and Topology, Rio de Janeiro, July 1976. Proceedings. Edited by J. Palis and M. do Carmo. VI, 866 pages. 1977.

Vol. 598: J. Hoffmann-Jørgensen, T. M. Liggett et J. Neveu, Ecole d'Eté de Probabilités de Saint-Flour VI – 1976. Edité par P.-L. Hennequin. XII, 447 pages. 1977.

Vol. 599: Complex Analysis, Kentucky 1976. Proceedings. Edited by J. D. Buckholtz and T. J. Suffridge. X, 159 pages. 1977.

Vol. 600: W. Stoll, Value Distribution on Parabolic Spaces. VIII, 216 pages. 1977.

Vol. 601: Modular Functions of one Variable V, Bonn 1976. Proceedings. Edited by J.-P. Serre and D. B. Zagier. VI, 294 pages. 1977.

Vol. 602: J. P. Brezin, Harmonic Analysis on Compact Solvmanifolds. VIII, 179 pages. 1977.

Vol. 603: B. Moishezon, Complex Surfaces and Connected Sums of Complex Projective Planes. IV, 234 pages. 1977.

Vol. 604: Banach Spaces of Analytic Functions, Kent, Ohio 1976. Proceedings. Edited by J. Baker, C. Cleaver and Joseph Diestel. VI, 141 pages. 1977.

Vol. 605: Sario et al., Classification Theory of Riemannian Manifolds. XX, 498 pages. 1977.

Vol. 606: Mathematical Aspects of Finite Element Methods. Proceedings 1975. Edited by I. Galligani and E. Magenes. VI, 362 pages. 1977.

Vol. 607: M. Métivier, Reelle und Vektorwertige Quasimartingale und die Theorie der Stochastischen Integration. X, 310 Seiten. 1977.

Vol. 608: Bigard et al., Groupes et Anneaux Réticulés. XIV, 334 pages. 1977.

Vol. 609: General Topology and Its Relations to Modern Analysis and Algebra IV. Proceedings 1976. Edited by J. Novák. XVIII, 225 pages. 1977.

Vol. 610: G. Jensen, Higher Order Contact of Submanifolds of Homogeneous Spaces. XII, 154 pages. 1977.

Vol. 611: M. Makkai and G. E. Reyes, First Order Categorical Logic. VIII, 301 pages. 1977.

Vol. 612: E. M. Kleinberg, Infinitary Combinatorics and the Axiom of Determinateness. VIII, 150 pages. 1977.

Vol. 613: E. Behrends et al., L^P-Structure in Real Banach Spaces. X, 108 pages. 1977.

Vol. 614: H. Yanagihara, Theory of Hopf Algebras Attached to Group Schemes. VIII, 308 pages. 1977.

Vol. 615: Turbulence Seminar, Proceedings 1976/77. Edited by P. Bernard and T. Ratiu. VI, 155 pages. 1977.

Vol. 616: Abelian Group Theory, 2nd New Mexico State University Conference, 1976. Proceedings. Edited by D. Arnold, R. Hunter and E. Walker. X, 423 pages. 1977.

Vol. 617: K. J. Devlin, The Axiom of Constructibility: A Guide for the Mathematician. VIII, 96 pages. 1977.

Vol. 618: I. I. Hirschman, Jr. and D. E. Hughes, Extreme Eigen Values of Toeplitz Operators. VI, 145 pages. 1977.

Vol. 619: Set Theory and Hierarchy Theory V, Bierutowice 1976. Edited by A. Lachlan, M. Srebrny, and A. Zarach. VIII, 358 pages. 1977.

Vol. 620: H. Popp, Moduli Theory and Classification Theory of Algebraic Varieties. VIII, 189 pages. 1977.

Vol. 621: Kauffman et al., The Deficiency Index Problem. VI, 112 pages. 1977.

Vol. 622: Combinatorial Mathematics V, Melbourne 1976. Proceedings. Edited by C. Little. VIII, 213 pages. 1977.

Vol. 623: I. Erdelyi and R. Lange, Spectral Decompositions on Banach Spaces. VIII, 122 pages. 1977.

Vol. 624: Y. Guivarc'h et al., Marches Aléatoires sur les Groupes de Lie. VIII, 292 pages. 1977.

Vol. 625: J. P. Alexander et al., Odd Order Group Actions and Witt Classification of Innerproducts. IV, 202 pages. 1977.

Vol. 626: Number Theory Day, New York 1976. Proceedings. Edited by M. B. Nathanson. VI, 241 pages. 1977.

Vol. 627: Modular Functions of One Variable VI, Bonn 1976. Proceedings. Edited by J.-P. Serre and D. B. Zagier. VI, 339 pages. 1977.

Vol. 628: H. J. Baues, Obstruction Theory on the Homotopy Classification of Maps. XII, 387 pages. 1977.

Vol. 629: W. A. Coppel, Dichotomies in Stability Theory. VI, 98 pages. 1978.

Vol. 630: Numerical Analysis, Proceedings, Biennial Conference, Dundee 1977. Edited by G. A. Watson. XII, 199 pages. 1978.

Vol. 631: Numerical Treatment of Differential Equations. Proceedings 1976. Edited by R. Bulirsch, R. D. Grigorieff, and J. Schröder. X, 219 pages. 1978.

Vol. 632: J.-F. Boutot, Schéma de Picard Local. X, 165 pages. 1978.

Vol. 633: N. R. Coleff and M. E. Herrera, Les Courants Résiduels Associés à une Forme Méromorphe. X, 211 pages. 1978.

Vol. 634: H. Kurke et al., Die Approximationseigenschaft lokaler Ringe. IV, 204 Seiten. 1978.

Vol. 635: T. Y. Lam, Serre's Conjecture. XVI, 227 pages. 1978.

Vol. 636: Journées de Statistique des Processus Stochastiques, Grenoble 1977, Proceedings. Edité par Didier Dacunha-Castelle et Bernard Van Cutsem. VII, 202 pages. 1978.

Vol. 637: W. B. Jurkat, Meromorphe Differentialgleichungen. VII, 194 Seiten. 1978.

Vol. 638: P. Shanahan, The Atiyah-Singer Index Theorem, An Introduction. V, 224 pages. 1978.

Vol. 639: N. Adasch et al., Topological Vector Spaces. V, 125 pages. 1978.

Vol. 700: Module Theory, Proceedings, 1977. Edited by C. Faith and S. Wiegand. X, 239 pages. 1979.

Vol. 701: Functional Analysis Methods in Numerical Analysis, Proceedings, 1977. Edited by M. Zuhair Nashed. VII, 333 pages. 1979.

Vol. 702: Yuri N. Bibikov, Local Theory of Nonlinear Analytic Ordinary Differential Equations. IX, 147 pages. 1979.

Vol. 703: Equadiff IV, Proceedings, 1977. Edited by J. Fábera. XIX, 441 pages. 1979.

Vol. 704: Computing Methods in Applied Sciences and Engineering, 1977, I. Proceedings, 1977. Edited by R. Glowinski and J. L. Lions. I, 391 pages. 1979.

Vol. 705: O. Forster und K. Knorr, Konstruktion verseller Familien kompakter komplexer Räume. VII, 141 Seiten. 1979.

Vol. 706: Probability Measures on Groups, Proceedings, 1978. Edited by H. Heyer. XIII, 348 pages. 1979.

Vol. 707: R. Zielke, Discontinuous Čebyšev Systems. VI, 111 pages. 1979.

Vol. 708: J. P. Jouanolou, Equations de Pfaff algébriques. V, 255 pages. 1979.

Vol. 709: Probability in Banach Spaces II. Proceedings, 1978. Edited by A. Beck. V, 205 pages. 1979.

Vol. 710: Séminaire Bourbaki vol. 1977/78, Exposés 507–524. IV, 328 pages. 1979.

Vol. 711: Asymptotic Analysis. Edited by F. Verhulst. V, 240 pages. 1979.

Vol. 712: Equations Différentielles et Systèmes de Pfaff dans le Champ Complexe. Edité par R. Gérard et J.-P. Ramis. V, 364 pages. 1979.

Vol. 713: Séminaire de Théorie du Potentiel, Paris No. 4. Edité par F. Hirsch et G. Mokobodzki. VII, 281 pages. 1979.

Vol. 714: J. Jacod, Calcul Stochastique et Problèmes de Martingales. X, 539 pages. 1979.

Vol. 715: Inder Bir S. Passi, Group Rings and Their Augmentation Ideals. VI, 137 pages. 1979.

Vol. 716: M. A. Scheunert, The Theory of Lie Superalgebras. X, 271 pages. 1979.

Vol. 717: Grosser, Bidualräume und Vervollständigungen von Banachmoduln. III, 209 pages. 1979.

Vol. 718: J. Ferrante and C. W. Rackoff, The Computational Complexity of Logical Theories. X, 243 pages. 1979.

Vol. 719: Categorial Topology, Proceedings, 1978. Edited by H. Herrlich and G. Preuß. XII, 420 pages. 1979.

Vol. 720: E. Dubinsky, The Structure of Nuclear Fréchet Spaces. V, 187 pages. 1979.

Vol. 721: Séminaire de Probabilités XIII. Proceedings, Strasbourg, 1977/78. Edité par C. Dellacherie, P. A. Meyer et M. Weil. VII, 647 pages. 1979.

Vol. 722: Topology of Low-Dimensional Manifolds. Proceedings, 1977. Edited by R. Fenn. VI, 154 pages. 1979.

Vol. 723: W. Brandal, Commutative Rings whose Finitely Generated Modules Decompose. II, 116 pages. 1979.

Vol. 724: D. Griffeath, Additive and Cancellative Interacting Particle Systems. V, 108 pages. 1979.

Vol. 725: Algèbres d'Opérateurs. Proceedings, 1978. Edité par P. de la Harpe. VII, 309 pages. 1979.

Vol. 726: Y.-C. Wong, Schwartz Spaces, Nuclear Spaces and Tensor Products. VI, 418 pages. 1979.

Vol. 727: Y. Saito, Spectral Representations for Schrödinger Operators With Long-Range Potentials. V, 149 pages. 1979.

Vol. 728: Non-Commutative Harmonic Analysis. Proceedings, 1978. Edited by J. Carmona and M. Vergne. V, 244 pages. 1979.

Vol. 729: Ergodic Theory. Proceedings, 1978. Edited by M. Denker and K. Jacobs. XII, 209 pages. 1979.

Vol. 730: Functional Differential Equations and Approximation of Fixed Points. Proceedings, 1978. Edited by H.-O. Peitgen and H.-O. Walther. XV, 503 pages. 1979.

Vol. 731: Y. Nakagami and M. Takesaki, Duality for Crossed Products of von Neumann Algebras. IX, 139 pages. 1979.

Vol. 732: Algebraic Geometry. Proceedings, 1978. Edited by K. Lønsted. IV, 658 pages. 1979.

Vol. 733: F. Bloom, Modern Differential Geometric Techniques in the Theory of Continuous Distributions of Dislocations. XII, 206 pages. 1979.

Vol. 734: Ring Theory, Waterloo, 1978. Proceedings, 1978. Edited by D. Handelman and J. Lawrence. XI, 352 pages. 1979.

Vol. 735: B. Aupetit, Propriétés Spectrales des Algèbres de Banach. XII, 192 pages. 1979.

Vol. 736: E. Behrends, M-Structure and the Banach-Stone Theorem. X, 217 pages. 1979.

Vol. 737: Volterra Equations. Proceedings 1978. Edited by S.-O. Londen and O. J. Staffans. VIII, 314 pages. 1979.

Vol. 738: P. E. Conner, Differentiable Periodic Maps. 2nd edition, IV, 181 pages. 1979.

Vol. 739: Analyse Harmonique sur les Groupes de Lie II. Proceedings, 1976-78. Edited by P. Eymard et al. VI, 646 pages. 1979.

Vol. 740: Séminaire d'Algèbre Paul Dubreil. Proceedings, 1977–78. Edited by M.-P. Malliavin. V, 456 pages. 1979.

Vol. 741: Algebraic Topology, Waterloo 1978. Proceedings. Edited by P. Hoffman and V. Snaith. XI, 655 pages. 1979.

Vol. 742: K. Clancey, Seminormal Operators. VII, 125 pages. 1979.

Vol. 743: Romanian-Finnish Seminar on Complex Analysis. Proceedings, 1976. Edited by C. Andreian Cazacu et al. XVI, 713 pages. 1979.

Vol. 744: I. Reiner and K. W. Roggenkamp, Integral Representations. VIII, 275 pages. 1979.

Vol. 745: D. K. Haley, Equational Compactness in Rings. III, 167 pages. 1979.

Vol. 746: P. Hoffman, τ-Rings and Wreath Product Representations. V, 148 pages. 1979.

Vol. 747: Complex Analysis, Joensuu 1978. Proceedings, 1978. Edited by I. Laine, O. Lehto and T. Sorvali. XV, 450 pages. 1979.

Vol. 748: Combinatorial Mathematics VI. Proceedings, 1978. Edited by A. F. Horadam and W. D. Wallis. IX, 206 pages. 1979.